Mastering Complexity

Stephen Denker

Mastering Complexity

Adding Coherence Throughout Your Business with Dependency Structure Spreadsheets

CRC Press
Taylor & Francis Group
Boca Raton   London   New York

CRC Press is an imprint of the
Taylor & Francis Group, an **informa** business

A PRODUCTIVITY PRESS BOOK

CRC Press
Taylor & Francis Group
6000 Broken Sound Parkway NW, Suite 300
Boca Raton, FL 33487-2742

© 2015 by Taylor & Francis Group, LLC
CRC Press is an imprint of Taylor & Francis Group, an Informa business

No claim to original U.S. Government works

Printed on acid-free paper
Version Date: 20141217

International Standard Book Number-13: 978-1-4987-0079-5 (Paperback)

This book contains information obtained from authentic and highly regarded sources. Reasonable efforts have been made to publish reliable data and information, but the author and publisher cannot assume responsibility for the validity of all materials or the consequences of their use. The authors and publishers have attempted to trace the copyright holders of all material reproduced in this publication and apologize to copyright holders if permission to publish in this form has not been obtained. If any copyright material has not been acknowledged please write and let us know so we may rectify in any future reprint.

Except as permitted under U.S. Copyright Law, no part of this book may be reprinted, reproduced, transmitted, or utilized in any form by any electronic, mechanical, or other means, now known or hereafter invented, including photocopying, microfilming, and recording, or in any information storage or retrieval system, without written permission from the publishers.

For permission to photocopy or use material electronically from this work, please access www.copyright.com (http://www.copyright.com/) or contact the Copyright Clearance Center, Inc. (CCC), 222 Rosewood Drive, Danvers, MA 01923, 978-750-8400. CCC is a not-for-profit organization that provides licenses and registration for a variety of users. For organizations that have been granted a photocopy license by the CCC, a separate system of payment has been arranged.

Trademark Notice: Product or corporate names may be trademarks or registered trademarks, and are used only for identification and explanation without intent to infringe.

Visit the Taylor & Francis Web site at
http://www.taylorandfrancis.com

and the CRC Press Web site at
http://www.crcpress.com

Contents

Preface ix
Acknowledgments xi

SECTION 1 Work Has Structure

Chapter 1 What Is Structure? 3

Chapter 2 Why a Structural Model? 9

Chapter 3 DSM Spreadsheet Process Model 13

Chapter 4 DSM Spreadsheet Shows a Plan 17

Chapter 5 Putting on My Shoes 19

Chapter 6 Reducing Cycle Time 23

Chapter 7 Reviews That Add Value 25

SECTION 2 Business Process Structures

Chapter 8 Visualizing a Business 31

Chapter 9 Cross-Organizational Collaboration 37

Chapter 10 Representing Complexity 43

SECTION 3 Building Dependency Maps

Chapter 11 What Depends on What?... 53
    11.1 Introduction 53
    11.2 Some Dependency Types..53
    11.3 Direct Dependencies 55

Chapter 12 Projects and Structure 59

Chapter 13 Making Assumptions.. 63

Chapter 14 Manipulating Assumptions... 67

Chapter 15 Active Risk Management.. 71

Chapter 16 Unwrapping Circuits 75

SECTION 4 Exposing Logical Structure

Chapter 17 Topological Order 81

Chapter 18 Partitioning 85

Chapter 19 Tearing Circuits.. 87
    19.1 Introduction 87
    19.2 What Tearing Does..89
    19.3 Problem-Solving Advice 89

Chapter 20 Structural Components 91

Chapter 21 Configuration Testing 93

Chapter 22 Control Dependencies..97

Chapter 23 Breakthrough Thinking 101

SECTION 5   Putting DSM Spreadsheets to Work

Chapter 24 Understanding Structure Essentials 109

SECTION 6   Tools

Chapter 25 Microsoft Excel®-Based Free Software.......................... 113

Chapter 26 Problematics PSM 32 117

Chapter 27 Lattix Architect and Lattix Analyst............................... 121

Index.. 123

About the Author.. 129

Preface

A business is a network of logical dependencies. Once the links representing dependencies are visualized in a spreadsheet, it can be used to improve the business. This book shows how to visualize these logical dependencies with a square spreadsheet called a *dependency structure matrix*, or *DSM*.

The goal of this book is to help develop and maintain logical, adaptable solutions to business problems. The focus is on simple practices and avoiding complex theories. To better help prepare for solving our own problems by ourselves, simple models are used because they are easy to understand.

- The best software matches the problem it is meant to solve and the way people are accustomed to working.
- Because we can use this software in familiar ways, it is easy to learn.

Spreadsheets first existed as paper ledgers for financial applications. Spreadsheet software combined graphics with computations to solve a wide range of problems. This software has proven so convenient we often adapt our problem-solving approach to a spreadsheet-like format. Today, spreadsheets are a fundamental template for organizing our thinking about a wide variety of problems.

Donald Steward[*] developed the simple, yet powerful, DSM tool. He called his first DSM a *design* structure matrix. Engineers still prefer this name.[†,‡] DSMs are applied to a much wider class of problems,[§] therefore using the term *dependency* structure matrix for DSM.

This book begins by identifying how logical *elements* are represented in a DSM spreadsheet. It shows where the elements can run in parallel, where the elements must run in sequence, and where the elements are tangled together. I show how to untangle these elements so they can be understood.

---

[*] Steward, Donald. 1981. *Systems Analysis and Management: Structure, Strategy, and Design.* New York: Petrocelli.

[†] Eppinger, Steven D. 2001. Innovation at the speed of information. *Harvard Business Review.* January: 149–158.

[‡] Eppinger, Steven D., and Browning, Tyson R. 2012. *Design Structure Matrix Methods and Applications.* Cambridge, MA: MIT Press.

[§] Clark, Kim, and Baldwin, Carliss. 1998. *Design Rules: The Power of Modularity.* Cambridge, MA: MIT Press.

Several powerful techniques are acquired to identify, analyze, rearrange, and deploy the logical elements of many complex business systems. One of the best features about these rearranged logical systems is that they have a tendency *to solve the system's problem in a way that it stays solved*. I hope DSM spreadsheets will enable you to solve many problems.

Acknowledgments

That this book even exists is in no small measure because of the critical inspiration and generous help I have received over many years from Don Steward and Jack Nevison. They are both colleagues and valued personal friends.

I have known and collaborated with Don since 1992. Don was the originator of the Dependency Structure Method and many of the collateral ideas. He is a typical creative thinker—self-centered, an explosive fountain of ideas, and idiosyncratic. He is highly ethical and a most rare resource. If it were not for his persistence and stubbornness, many of his great ideas would have died on the vine years ago.

I first met Don Steward at the Massachusetts Institute of Technology Sloan School, where he was spending his sabbatical leave from California State University, Sacramento. His enthusiasm and creativity are contagious. I heard his ideas and became an acolyte, spending the years since that day understanding, extending, teaching, and transferring the concepts embodied in his new idea he called the Design Structure Matrix or DSM.

Don wrote his first paper in 1968 on his new concept while at General Electric. Like many innovative thinking paradigms and tools, the paper was rejected for publication and was not finally accepted until 1981.[*]

In 1981, Don also published a book—*Systems Analysis and Management: Structure, Strategy, and Design* (New York: Petrocelli)—to explain his method in depth and give many examples drawn not only from engineering and product design but also from social problems and nontechnical situations. His book has long been out of print; I have undertaken the writing of my book in part to continue Don's initiatives.

Currently, Don is emeritus professor at California State University, Sacramento, and a principal at Problematics, LLC. In 1987, he also wrote another book, *Software Engineering with Systems Analysis and Design* (Pacific Grove, CA: Brooks/Cole).

I met Jack Nevison later, and we also have become collaborators. Jack was the person who encouraged me to write this book. He spent many

---

[*] Steward, D. V. 1981. The design structure system—A method for managing the design of complex systems. *IEEE Transactions on Engineering Management* EM-28 (August): 71–74.

hours reading, critiquing, and helping me strengthen its content clarity and exposition.

Jack himself is the author of six books and numerous articles on computing and management. During the course of his business career, he built and sold several businesses, managed projects, managed project managers, and served as both an internal and external consultant to Global 1000 companies.

There are many others who over the years have helped me better understand the DSM spreadsheet concept and its applications and great potential. Professors Steven Eppinger and Tyson Browning have expanded the application scope of the DSM spreadsheet and its worldwide awareness, especially among the engineering community. Today, DSM spreadsheets are used in many countries in both Europe and Asia. There is a continuous flow of innovative articles and publications by their students and colleagues.

I have also benefited by working on the problems my clients have provided me and in our many conversations. In particular, I want to thank Frank Waldman, Joseph Killough, Rich Payne, Hugh McLaughlin, Simon Austin, and Nicolay Worren.

Although I take full responsibility for any errors or confusion in this book's material, it would not be possible to write it at all without their input and support.

Here is Edward Bear coming downstairs—bump, bump, bump on the back of his head behind Christopher Robin. It is, as far as he knows, the only way of coming downstairs, but sometimes he feels that there really is another way, if only he could stop bumping for a moment and think of it.

—**A. A. Milne,** *Winnie the Pooh*

Section 1

Work Has Structure

By relieving the mind of all unnecessary work, a good notation sets it free to concentrate on more advanced problems, and in effect increases the mental power of the human race.

—**Alfred North Whitehead**

1

What Is Structure?

Does our organization functional structure match the logical dependency among its elements? If not congruent, it costs us time and money. If we visualize our dependency structures, we can improve our business or know better how to cope.

Some think the only structure in a business is its organization chart. An organization chart is a valuable administrative convenience. But, it should not be confused with the *what*, *why*, and *how* of the business. All too often, it is the organization chart, not the business, that is managed (see Figure 1.1).

Others think structure means the design of organizational workflows and business processes.

- In this book, *structure* is any pattern of logical relationships among key components.

What we call *structure* shows how elements are logically connected together. Structure shows whether one part needs another. The meaning of why they do is called *semantics*. Together, they describe our problem.

- We make this distinction so we can extract and study the dependency structure of a problem separate from its meaning.

Every business has these structures. They might include our whole organization, its employee attitudes, the product quality, the ways decisions are made, or hundreds of other factors. The word *structure* comes from the Latin *struere*, "to build." Dependency structures are built from the choices we make—things inside or outside our organization or *anything that matters anywhere*. Structures are built from the choices we make over time.

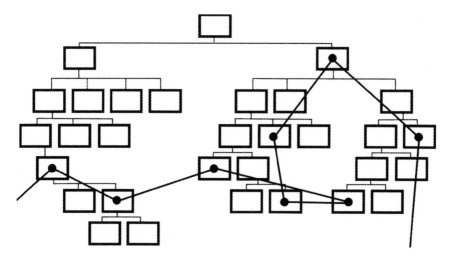

FIGURE 1.1
How dependencies travel throughout an organization and beyond.

When we ask, What happens if x changes? we begin to see how an element might be part of a dependency structure. Dependency structures occur everywhere.

- If we can make them visible, the implications are enormous.

Actions cannot be considered independently. If we are used to thinking in terms of tasks rather than their underlying dependencies, we may need to recast our point of view to move from a *logistics view* (managing tasks) to a *structural view* (managing dependencies).

Every business consists of networks of dependency structures, of persons and their dependent activities. Why they do what they do is called *semantics*. Together, structure and semantics describe our work.

- The *structure* and its *semantics* determine how our business performs.

As dependency structures become larger, diagrams drawn to illustrate relationships can become cluttered with arrows and connecting lines.

"To assemble an annual financial plan for its 3,500 U.S. stores, Sears collected data scattered across many computers. It took a *100-square-foot diagram* [emphasis added] *to describe their 300-step consolidation process*" (*Business Week*, October 29, 1996: 131). A simple DSM spreadsheet model could have captured all the relevant paths in one clear spreadsheet.

A DSM spreadsheet concisely visualizing both entire enterprises and their local connections *on one page* would clearly show how one thing triggers a cascade.

Figures 1.2 and 1.3 are other examples of how complexity that could overwhelm our ability to comprehend is made more manageable when converted to a DSM spreadsheet visualization. The process of how to accomplish this is illustrated in the chapters that follow.

Our client, in preparation for reworking data entry capabilities in anticipation of possible Y2K (year 2000) computer errors as the twentieth century was ending, diagrammed their enterprise as shown in Figure 1.2. I joined the project. Our first task was to draw the equivalent DSM spreadsheet shown in Figure 1.3. With the DSM spreadsheet visualization complete, immediately our client saw three errors in their model.

- Correcting their graph took some time.
- Correcting the DSM spreadsheet visualization took less than one minute.

With DSM spreadsheets, we can show clearly how individuals are connected. We can understand how their relationships are dependent. More important, we can systematically revise and optimize these dependencies one at a time or as a whole.

In this book, we first treat simple structures. As illustrated problems become more complex, we introduce the tools we need. With these tools, we can consider the concepts as applied to larger systems.

Let us get started!

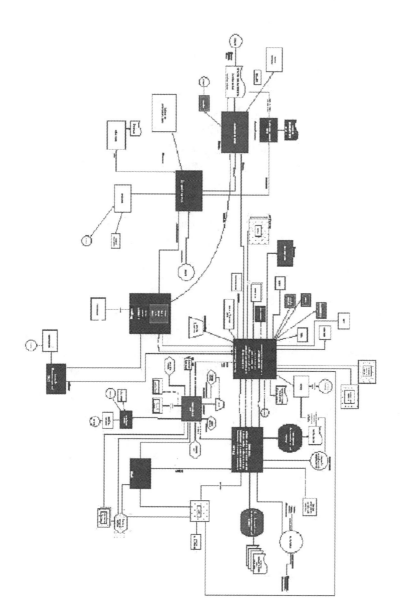

FIGURE 1.2
Original Y2K enterprise diagram drawn as a graph.

Top

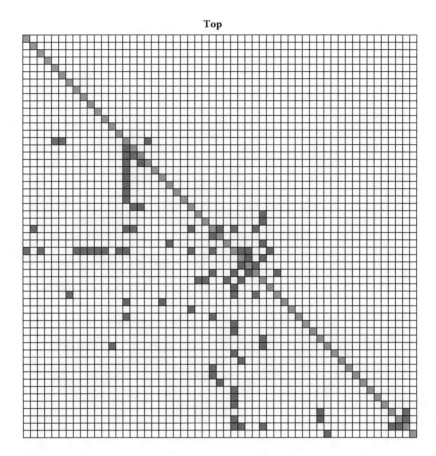

FIGURE 1.3
Original Y2K enterprise diagram drawn as a DSM spreadsheet.

2
Why a Structural Model?

How can we know that a proposed organizational change will be an organizational improvement?

- By looking at the organization's logical structure.

Using a DSM spreadsheet, we can make this structure visible and its key logical elements self-evident. A DSM spreadsheet helps us articulate, verify, and reorganize structures.

- We use the general term *element* to denote a row (and its column) in a DSM spreadsheet.

In a DSM spreadsheet, each element will be represented twice—once as a row and again as a column. Elements are listed down the left-hand side of the DSM spreadsheet. The same list of elements is placed along the top of the DSM spreadsheet, from left to right. Their row order and their column order are identical. Each element's row and column intersect at the DSM spreadsheet diagonal cell.

- Elements in a DSM spreadsheet can represent many entities: tasks, physical objects, time, human resources, a person's point of view—*anything*. And, they need not all be the same type.

Also, we sometimes refer to DSM spreadsheet elements as *nodes*. A *graph* is a set of points, or nodes, and the lines between them. When the lines have a direction, they are called *arcs* in a *directed graph*. *A DSM spreadsheet and directed graph are alternative, equivalent ways to map dependencies*. A sequence of arcs from node to node is a *path*.

10 • *Mastering Complexity*

Some nodes might be arranged in *parallel paths* because they do not depend on each other. Others might be arranged in *sequence* because each depends on the results of one before.

If a path eventually returns to a node, the path is a *circuit*. Circuits exist because elements can jointly depend on each other. But, time cannot abide circuits. *There is no logical place to start.* Figure 2.1 displays a DSM spreadsheet and its equivalent graph, with one of its circuits identified.

A large number of problems we face in business contain logical circuits of dependencies. For example, if we are developing a new product, marketing needs to know what it is going to be so a strategy can be planned. Manufacturing needs plans to build it. Engineering needs to know what to make. Marketing needs to supply this information, but to do so, costs are needed. Engineering asks manufacturing, which replies, "Well, tell us what the design is and how many we are going to make."

- We find ourselves locked in a logical circuit—in a loop of highly dependent tasks.
- With a DSM spreadsheet, we can recognize logical circuits, the first step in solving an organizational problem.

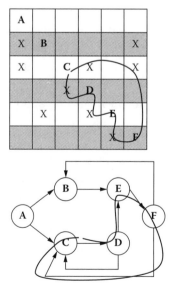

FIGURE 2.1
A logical circuit represented as a graph and as a DSM spreadsheet.

We need to resolve what we are going to do with circuits. Chapter 8 provides an example showing how to do this.

A DSM spreadsheet can help us solve complex problems where there are many related elements.

- DSM spreadsheet elements can be rearranged in a logical way.

We can break apart circuits. We can unwrap them. We can order them in time. We can convert circuits into sequences that can be scheduled using conventional techniques. Chapter 16 shows ways to do this.

3
DSM Spreadsheet Process Model

Each *element* of a DSM spreadsheet is assigned to a *row* and its corresponding *column* in a DSM spreadsheet. A DSM spreadsheet *cell* is the intersection of a row and a column.

- ***Rows and columns are ordered identically. Marks in a DSM spreadsheet cell show where one element depends on another. A mark in a row shows that a directed relationship to the row element from a column element exists.***

As marked, a DSM spreadsheet defines an equivalent directed graph whose arcs between nodes correspond to marks in the DSM spreadsheet. The two elements at the intersection are related to each other. Arcs entering a graph node are from *predecessor* nodes, and arcs leaving a node go to *successor* nodes. So, in a sense in a DSM spreadsheet, a mark in the cell at the intersection of row A and column B represents the graph arrowhead of the arc from node B to node A (B precedes A). An empty cell indicates no link—the two elements are independent of each other. See Figure 3.1 for examples of the three possible ways two elements can be connected structurally.

Figure 3.2 shows a more complex example of a graph and its equivalent DSM spreadsheet having the same dependencies. Chapter 4 revisits this example. We will rearrange the DSM spreadsheet elements.

- **In rearranging a DSM spreadsheet, we move both a row and its corresponding column together the same amount in the same way so the intersecting spreadsheet cell remains on the diagonal.**

This is why we use the term *element* for a row or for a column. We use the words *row* and *column* only when we mean specifically a particular row

14 • *Mastering Complexity*

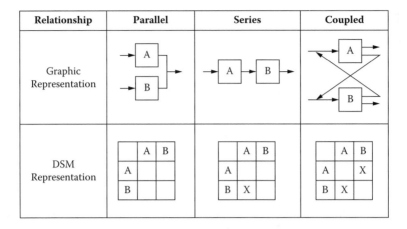

FIGURE 3.1
Two-element graphs and their equivalent DSM spreadsheet models.

or a column. Dependencies remain unchanged, no matter the order of the rows and columns.

- **So, how do we answer the tough question, what do we owe others so they can do their job?**

We know what we need. But, we might not be entirely sure who needs our work—who needs what we generate. We can organize both what we need and what we owe. From a microlevel view, we can obtain the required macrolevel view. But, only I can answer the question: What do I need? From whom? In what form? In what time frame?

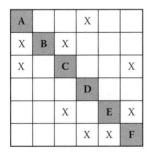 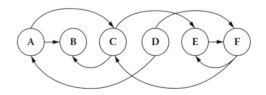

A needs something from **D**
B needs something from **A**, and **C**
C needs something from **A**, and **F**
D needs nothing
E needs something from **C**, and **F**
F needs something from **D**, and **E**

FIGURE 3.2
A larger graph and its equivalent DSM spreadsheet showing the same dependencies.

- *Using a DSM spreadsheet, one question—our only question—is what do we need?*

We build up the DSM spreadsheet row by row, so we have all the columns as well. This is because once we build our very first DSM spreadsheet, the columns tell us where everything is needed by others.

- **Regardless of the order of rows, we can know who needs our work.**

- *Often, we do not need to do anything more. Once we each have successively defined the individual needs of every row, DSM spreadsheet columns automatically provide the answers to the question, What do we owe?*

This information creates our links to others and to the organization. It enables us to do our job.

We can understand both what we need and what we owe. Then, we can set about organizing both.

- *We start from a local view, and we obtain a global view.*

- *We can see what we all need.*

By going from a micro- to a macroview, we can organize what we all need and then collectively focus on action.

- **The purpose of information is not just knowledge. It is the ability to take the right actions.**

Both questions, What do I owe? and What do I need? sound deceptively simple. It takes a lot of thought and work to answer them, and our answers are not forever. These questions have to be asked again every time there is a real change.

- **For each DSM spreadsheet row, we only identify an element's direct dependencies. We identify the elements on which it is directly dependent.**

- We construct a DSM spreadsheet row by row. For each row, we simply move left to right, putting marks in those cells where there are dependencies.
- Marks in rows show where we need something. Reading down a column reveals where that element owes something to a row element.

4
DSM Spreadsheet Shows a Plan

A DSM spreadsheet can be organized to show which elements can be done in parallel, which elements must be done in sequence, and which elements are tangled together.

The rows of the DSM spreadsheet in Figure 4.1 can be rearranged so dependencies flow forward down the page. This puts most marks below the diagonal, as shown in Figure 4.2. See the chapters in Part 4 of this book for how we do this.

If there are some elements that depend on one another, it is not possible to order a DSM spreadsheet so all marks are below the diagonal. Grouped together, these elements form a square block astride the diagonal as shown in Figure 4.2. Within a block, there are marks both above and below the diagonal. Change any one and all the rest in that block will be affected. If we pick any node, we can follow arcs to obtain to any other node in that block and back again (i.e., a circuit) (see Figure 2.1).

- Time has no role in these dependencies.
- This circular property is what prevents us from knowing what to do first.
- Within a block, there is no place to start unless we begin by making some estimates or assumptions.

A DSM spreadsheet is a vivid way of spotting the existence of circuits. Long iteration loops are rarely efficient. Because they occur in large blocks, they imply decisions reached later can force us to reevaluate many steps occurring much earlier. The quicker iteration takes place the better. We can make any corrections before much time has passed. These quicker iterations occur in smaller blocks.

We try to rearrange elements so any iteration loops are in as small blocks as possible. So, how do we do this? Read on to find the answer.

18 • *Mastering Complexity*

	A			X	
X	**B**	X			
X		**C**			X
			D		
		X		**E**	X
				X	X	**F**

A needs something from **D**
B needs something from **A** and **C**
C needs something from **A** and **F**
D needs nothing
E needs something from **C** and **F**
F needs something from **D** and **E**

FIGURE 4.1
DSM spreadsheet as originally drawn. Note that there are unnecessary iteration loops—feedbacks that would add no value.

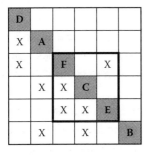
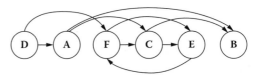

(Note: Whenever an element appears, its dependencies remain the same.)

D is first - it needs nothing
A depends on **D** and **D** has occurred
F depends on **D** and (**E** has not occurred yet)

Rows **C**, **E**, and **B** depend on elements that have occurred
B is last, since it feeds nothing

FIGURE 4.2
Reordering the DSM spreadsheet removes unnecessary iteration loops.

5

Putting on My Shoes

Figure 5.1 shows a DSM spreadsheet for a familiar process, one we do every day—putting on our shoes. The marks below the DSM spreadsheet diagonal indicate, for instance, *Get Socks* must precede *Put on Socks*, and *Get Shoes* must precede *Put on Shoes* and *Inspect Shoes*. Reading across row 3, we see *Put on Socks* depends on *Get Socks*; reading across row 4, we see that *Put on Shoes* depends on *Get Shoes* and *Put on Socks*. Reading across row 2, we see *Get Shoes* depends on *Inspect Shoes*. Because this mark is above the diagonal, we make an assumption about this dependency and proceed without it. But, the mark above the DSM spreadsheet diagonal indicates that once we have inspected our shoes, we might find that they are scuffed or the wrong color, requiring an iterative circuit, *Get (New) Shoes*.

Our goal is resequencing tasks to (1) remove iterations or (2) minimize their impact.

- The tasks *Get Shoes* and *Inspect Shoes* are coupled through a feedback loop.
- There is no way to remove iteration by reordering the rows and columns to have all the marks entirely below the DSM spreadsheet diagonal.

So, we change our goal to *shorten any iteration* by getting any marks in as small blocks as possible. In Figure 5.1, once we *Get Shoes*, we *Put on Socks* and *Put on Shoes* before we *Inspect Shoes*. If, instead, we move inspection upstream as shown in Figure 5.2, we minimize the impact of any need to get (new) shoes. Although in this case we could not eliminate iteration entirely, often it is possible to reduce the length of the iteration loops substantially by resequencing rows and columns.

20 • *Mastering Complexity*

	Get Socks	Get Shoes	Put on Socks	Put on Shoes	Inspect on Shoes
Get Socks	1				
Get Shoes		2			X
Put on Socks	X		3		
Put on Shoes		X	X	4	
Inspect Shoes		X			5

FIGURE 5.1
DSM spreadsheet for putting on my shoes.

The DSM spreadsheet also indicates which tasks can be accomplished in parallel—there is no mutual dependency. For example, in the figures, *Get Socks* and *Get Shoes* can be done simultaneously, as can *Inspect Shoes* and *Put on Socks* (if we have enough resources—someone—to help us).

- Doing things concurrently is often seen as a way to reduce cycle time.
- If we choose to do tasks in parallel without considering their dependencies, this can lead to unnecessary rework and *increased* cycle time.

Arbitrarily doing tasks concurrently does not guarantee reduced cycle time.

	Get Shoes	Inspect Shoes	Get Socks	Put on Socks	Put on Shoes
Get Shoes	1	X			
Inspect Shoes	X	2			
Get Socks			3		
Put on Socks			X	4	
Put on Shoes	X			X	5

FIGURE 5.2
A better way of putting on my shoes.

Tasks OK to perform in parallel might be held up by other dependencies. The DSM spreadsheet indicates that *Get Socks* and *Get Shoes* can occur in parallel. But, if I am the only one available for both tasks, I might not be able to do both at once. A more complete analysis must account for these constraints.

6
Reducing Cycle Time

When we plan projects, we are likely to *hardwire* a particular ordering of tasks—what happens sequentially and what occurs concurrently. We describe the way we have always done things (*which probably needs to be improved*) or the way we would like things to be done (*which might not be feasible*).

- If we are used to thinking only in terms of tasks rather than their underlying dependencies, we need to recast our point of view. In projects, we move from a logistics view (*managing tasks*) to a structural view (*managing dependencies*).

For example, relationships between *design* elements describe a product. To design that product, we undertake tasks to set final values for the elements. We carry out these tasks over time. But, the dependencies between elements mean their final values all must be satisfied *simultaneously* even though they are determined *sequentially*. This often requires iteration.

- Using dependencies rather than tasks, we can know what questions to ask and how to arrange our answers.

We need a process description for both planning and execution. Analyzing the dependency structure can provide a more efficient and predictable process. Effective exchange of information is key to coordinating collaborative development. Accomplishing this exchange is fundamental. Our final DSM spreadsheet serves as a template for the workflow.

- A DSM spreadsheet helps us order tasks based on managing dependencies.

Let us illustrate this with an example: A software company needed to meet aggressive, fixed delivery requirements. It provided product enhancements to service-contract customers on a 6-month recurring basis. Each cycle, they shipped a package containing such items as bug fixes, version releases, upgrades, and new documentation.

Unfortunately their design, test, and production processes required over 80 steps and an 18-month completion cycle. Using project schedule charts, teams met over several months to try to reduce the 18 months to 6 months. Their objective was to keep their creation-fulfillment process cycle synchronized with their customer commitment cycle.

Team members were too close to the existing process to see how to rethink the task sequence. They made little progress. So, they decided to create a DSM spreadsheet. They started with the original task list. Then, they threw out their old schedule's workflow predecessor–successor dependency logic altogether. Instead, they interviewed key players, asking dependency questions: "What is essential to make our decision?" "What information do we need to start, update, or finish our work?" This made the complete problem clear. With this new dependency structure visible, they quickly saw several ways of immediately reducing an 18-month schedule at no risk. They accomplished their objective of 6-month cycle times.

The vast majority of the improvement came from recognizing which dependencies drove the workflow. To achieve the 67% cycle reduction time, they safely eliminated or realigned tasks that had been unrecognized sources of rework. They wasted less time by not doing tasks that added no value and reduced rework iterations caused by out-of-order tasks. They found low-risk opportunities to increase concurrency.

Rearranging tasks also streamlined the review process. See the next chapter for one of the ways they did it.

7
Reviews That Add Value

A DSM spreadsheet can be a powerful tool when we are planning a project life cycle—what occurs in what phases and where management reviews are necessary. We can schedule reviews so they occur when they add the most value.

Many projects are organized in groups of tasks or major phases. When project reviews are scheduled at the end of each phase, verification becomes the gateway to entering the next phase. Whenever verification is a gateway to entering the next phase, a project can be locked into an unnecessarily lengthened schedule.

A DSM spreadsheet can show us how to bypass this lock—how to reduce the number of reviews or remove them altogether. But first, we need to understand the DSM spreadsheet flow or we may ask the wrong questions.

What we do in what order establishes what assumptions we must make. Each sequence of the rows and their corresponding columns in a DSM spreadsheet corresponds to an approach. Whenever an approach is changed, dependencies remain the same. This gives us an opportunity to play with various approaches and their different assumptions to find opportunities to save time through increased concurrency and specifically tailored reviews.

We must perform a number of tasks before we can finally determine the consequences of those assumptions and decide whether they were suitable. What we do in what order depends on the assumptions we make and vice versa. Because our assumptions are made explicit, we can see where they need to be verified and understand how our plan must change if they prove to be inadequate. Only then would it be appropriate to formally review the tasks we have completed so far. Some tasks will have to be iterated. After we confirm the assumptions made, we are finished.

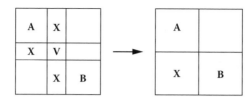

FIGURE 7.1
DSM spreadsheets showing verification and its removal.

For example, in the left DSM spreadsheet in Figure 7.1, **V** is a verification review for block **A**. Block **B** is dependent only on **V**, and **V** depends only on block **A**. Here, **X** represents these dependencies.

- We begin by removing all review tasks in our DSM spreadsheet so it only includes work tasks.
- Afterward, we will put back verification reviews tailored to the dependency structure.

Now, it becomes possible to have fewer, targeted project phase reviews and significantly increase concurrency.

In the right DSM spreadsheet of Figure 7.1, we have eliminated **V** and show only a direct dependency of block **B** on block **A**.

We can then see where concurrency might be possible. Let us look at the DSM spreadsheets in Figure 7.2.

If we consider just the elements that couple blocks **A** and **B**, we can group them into a new block, **C**.

Elements in blocks **A** and **B** are coupled by the elements in block **C**. This first DSM spreadsheet can be reordered to obtain the middle DSM spreadsheet. Block **C** sets the requirements for both **A** and **B**. Once the tasks in block **C** are carried out, tasks in blocks **A** and **B** can be carried out in parallel.

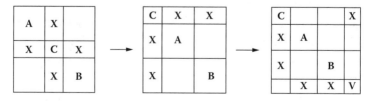

FIGURE 7.2
DSM spreadsheets suggesting better placement of verification.

But, blocks **A** and **B** are coupled by feedback through block **C**. To deal with this, we now insert a final verification review **V**, replacing any intermediate feedback, as shown by the right-most DSM spreadsheet. When the last task in block **B** is completed, review **V** can establish whether our assumptions are valid or whether another iteration of some or all of the tasks in the block is required. If another iteration is made, the schedule would be modified accordingly. See Chapter 16 for an example of one way we can do this.

The net effect is more concurrency and less-frequent and less-restrictive reviews, especially for product design projects.[*]

Working through large blocks usually involves more tasks and more people. So, our strategy is, if possible, to subdivide any large blocks into arrays of smaller ones. When there are small blocks, we are more likely to communicate more. Chapter 8 shows how this might work.

---

[*] This strategy is explained at great length in Clark, Kim, and Baldwin, Carliss. 1998. *Design Rules: The Power of Modularity*. Cambridge, MA: MIT Press.

Section 2

Business Process Structures

No matter how complicated a problem is, it usually can be reduced to a simpler, clearer form that is often the best solution.

—**An Wang**

8

Visualizing a Business[*]

In this chapter, a DSM spreadsheet is used to represent business process elements, map their dependencies, and design alternatives.

To enable business units to work together we *need* to understand how they *depend* on each other. Obviously, engineering, manufacturing, and marketing cannot all occur at the same time. If we forget time and tasks for a moment and just consider the relationship of *needs*, we can actually map the problem.

- STEP 1: IDENTIFY ELEMENTS

For step 1, the elements of the process need to be identified. We want to capture the main categories at the right level of detail. This depends on our needs. Each element can be as detailed as one person's tasks or as general as the responsibility of one business unit. Let us organize a product development process into the following elements:

A	Create Business Case
B	Develop Manufacturing Capacity
C	Analyze Technical Feasibility
D	Refine Product Concept
E	Determine Sourcing Strategy
F	Conduct Market Test

---

[*] This example follows an unpublished work by Nicolay Worren.

32 • *Mastering Complexity*

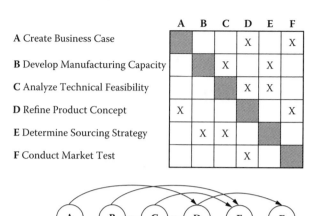

FIGURE 8.1
DSM spreadsheet and its equivalent graph of tasks.

- STEP 2: MAP DEPENDENCIES

During step 2, we determine dependencies. Different approaches might be taken. The best approach is a participative one—representatives from the business units involved discuss how their needs relate. Once created, DSM spreadsheets can become a shared resource for referral when carrying out work.

The graph and DSM spreadsheets in Figure 8.1 both show the same dependency relationships and provide alternative ways of illustrating the process structure caused by these dependencies.

A *Create Business Case* requires information from **D** *Refine Product Concept* and **F** *Conduct Market Test*. This might be because the business needs to quantify expectations for revenue and costs on a precise definition of the product and its expected market size.

- STEP 3: DESIGN PROCESS

If there are any marks above the diagonal, as in this case, we can try to improve both cycle time and cost by resequencing elements. Figure 8.2 illustrates one resequencing. Now, we can see our business requirements more clearly. This enables us to design a business organization to support them. There are coupled tasks. So, we were not able to get all marks below the DSM spreadsheet diagonal, but we can get them as close as possible.

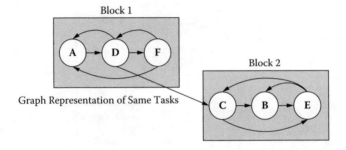

FIGURE 8.2
Rearranged tasks.

This creates blocks of closely related tasks so that iterations only occur between tasks that are near each other in the sequence.

The one exception is **D** and **C**, which belong to different blocks but are coupled. In Figure 8.2, two blocks of coupled tasks are connected through an information exchange between tasks **D** and **C**. We might also want to look more closely at tasks within these blocks (especially if the block consists of a large number of tasks).

- STEP 4: ALIGN ORGANIZATION

Figure 8.3 shows one organization of business units (step 4, align organization) that could operate under the dependency constraints identified.

- OPTION: USE SOME DESIGN RULES

The DSM spreadsheets can be used in a more fundamental way—to enable arrangements of organizational units so they could work independently

34 • *Mastering Complexity*

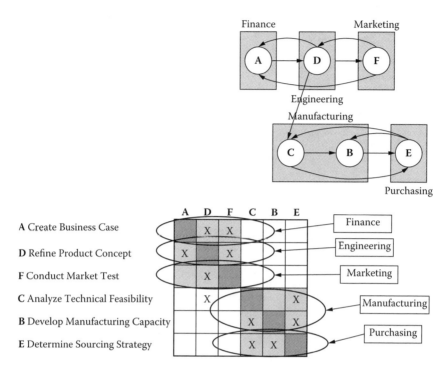

FIGURE 8.3
One organizational design.

yet achieve common business goals. If our process is going to be repeated often, or our DSM spreadsheet maps indicate a large number of iterations, we could create a general rule (called a *Design Rule* in product development). A *Design Rule* might say that either our development process will make use of known components or designers will develop new products to agreed standards. **C** *Analyze Technical Feasibility* requires information from **D** *Refine Product Concept*. *Analyze Technical Feasibility* might be considerably simplified, and we can proceed based on knowledge of the overall product concept alone, without waiting for *Refine Product Concept*. This entirely removes the direct coupling between phases.

By using *Design Rules*, we would reduce the need for ongoing coordination. This makes it possible for two process phases to proceed in parallel, as indicated in Figure 8.4.

Note that our case is simplified in several ways. Among other things, it shows only one *Design Rule*. This is consistent with the fact there was only one dependency between the blocks. There will usually be many more.

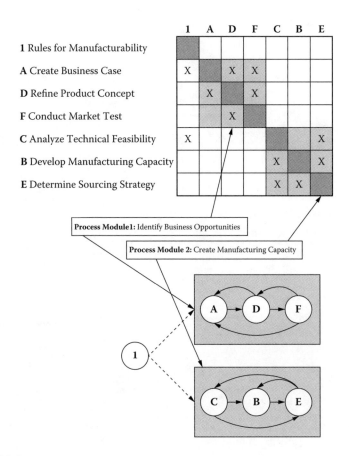

FIGURE 8.4
Design rule-based organization.

9

Cross-Organizational Collaboration

Businesses are organized to facilitate solving what is believed to be the principal goals for the organization. However, we must deal with problems residing within multiple dependency structures. Their intrinsic dependency structures constrain effective solutions. But, there may be trade-offs. If functional integrity is deemed more essential than a specific project's lead time or cost, it might be preferable to keep the current organizational structure but carefully manage cross-functional exchanges.

Business processes are the flows of work and information throughout our business. Some processes (e.g., software programming) might be contained wholly within a function. Most processes (e.g., order fulfillment) are cross functional, spanning between organizations.

What is the relationship between information flows and organizational structure? Each has its own structure of flows. Within our own group, interactions are frequent. We build and renew knowledge within our own group. In contrast, interactions across groups seem less important. We may forget knowledge crosses over group boundaries (see Figure 1.1 in Chapter 1).

This example deals with three critical, but different, views of the same process at a large automobile company[*]: (1) tasks, (2) responsibilities, and (3) communications.

Work was to be carried out sequentially, but by planned handoffs between several company groups. Five separate corporate groups were to be involved. Many information transfers between groups would occur.

---

[*] This example is adapted from two Massachusetts Institute of Technology Sloan School of Management theses: McCord, Kent, Managing the Integration Problem in Concurrent Engineering, 1993; and Dong, Qi, Representing Information Flow and Knowledge Management in Product Design Using the Design Structure Matrix, 1999.

The DSM spreadsheet in Figure 9.1 represents the project's entirely sequential development tasks.

The simplest situation is an organization whose functional structure matches logical dependency among elements. If these structures are not the same, it costs time and money (see Figure 9.2). If structures are made explicit, we can design improved business processes or know better how to cope.

There are trade-offs. If functional excellence is deemed more critical than project lead time or cost, it might be preferable to keep the current organizational structure but add a person to specifically manage the cross-functional process (see Figure 9.3).

The 2-year project was to develop the next generation of manufacturing lines. Built in were corporate requirements that *critical tasks* were to be performed in different departments. This created the need for significant coordination across formal boundaries. Resources could not be shuffled freely. The needed skills were not always available for the asking. But, they were not to colocate team members to facilitate communication. This created the need for significant coordination across functional boundaries.

Three DSM spreadsheets were used to visualize and identify critical needs for integration and project cross-functional coordination. Most of the tasks were easily scheduled sequentially.

- TASKS

Although the tasks shown in Figure 9.1 were sequential, two additional DSM spreadsheets were used to visualize the implementation plan and cross-functional coordination. Together, three DSM spreadsheet visualizations were used to help management resolve identified concerns. This DSM spreadsheet represents the sequence of tasks.

- RESPONSIBILITIES

Five separate corporate groups were involved, as shown in Figure 9.2. Special project managers were required to facilitate transfer and handoff of technical information from group to group.

Figure 9.2 shows the large amount of information transfer between groups. Although tasks were sequential, the plan as organized was not. Handoffs were likely to be problematic. This caused management to ask, "Have we established sufficient mechanisms to facilitate this information

	Preliminary business case	Identify needs & objectives	Establish targets	Generate selection criteria	Preliminary benefits/cost estimates	Prioritize alternatives	Select development direction	Identify high risk failure modes	Approval of funding & resources	Initiate concept validation	Engineering evaluation	Document engineering evaluation	Assess risks based on evaluation	Refine cost estimates	Identify candidate applications	Develop rollout plan	Secure divisional & corp.	Design prototype hardware	Build prototype hardware	Conduct validation testing	Design production hardware	Build production hardware
Preliminary business case	■																					
Identify needs & objectives	X	■																				
Establish targets	X	X	■																			
Generate selection criteria	X	X	X	■																		
Preliminary benefits/cost estimates			X	X	■																	
Prioritize alternatives				X	X	■																
Select development direction		X	X	X	X	X	■															
Identify high risk failure modes					X		X	■														
Approval of funding & resources	X					X			■													
Initiate concept validation						X		X		■												
Engineering evaluation		X	X			X	X	X	X		■											
Document engineering evaluation											X	■										
Assess risks based on evaluation								X			X	X	■									
Refine cost estimates				X							X			■								
Identify candidate applications											X		X		■							
Develop rollout plan											X		X		X	■						
Secure divisional & corp.															X	X	■					
Design prototype hardware											X		X			X	X	■				
Build prototype hardware																		X	■			
Conduct validation testing																X				■		
Design production hardware								X										X		X	■	
Build production hardware																					X	■

FIGURE 9.1
DSM spreadsheet of product development tasks.

Advanced Engineering	Preliminary Business Case	1																					
	Identify Needs & Objectives		2																				
	Estimate Benefits & Costs			5				X	X														
	Prioritize Alternatives	X		X	6				X														
	Initiate Concept Validation					10				X		X											
	Design Prototype Hardware						18						X	X				X	X				
Steering Group	Establish Targets	X	X					3															
	Generate Selection Criteria	X	X					X	4														
	Select Development Direction		X	X	X			X	X	7													
	Identify High-risk Failure Modes			X						X	8												
	Allocate Funding & Resources		X							X		9											
	Assess Risks Based on Evaluation										X		13	X	X								
Design Audit Team	Conduct Engineering Evaluation	X	X			X		X		X	X	X		11									
	Document Engineering Evaluation													X	12								
	Refine Cost Estimates			X										X		14							
Development Group	Select Application Site												X	X			15						
	Develop Rollout Plan												X	X			X	16					
	Secure Divisional Sign-off																X	X	17				
	Build Prototype Hardware						X													19			
	Conduct Validation Testing																	X			20		
Product Team	Design Production Hardware						X							X							X	21	
	Build Production Hardware																					X	22

FIGURE 9.2
DSM spreadsheet of project tasks reordered by responsible corporate group.

flow?" This prompted management to reexamine how information would be transferred among groups to answer the question.

It was discovered that only a few engineers were designated to coordinate required information exchange between groups. Figure 9.3 shows the responsible engineers' names within the project.

- COMMUNICATIONS

One manager, who we will call Tiger, was the key link between the *Advanced Engineering Team* and several other groups. But, Tiger was overextended. Tiger was both the lead engineer of the *Advanced Engineering Team* and a key member of the *Project Steering Team*. Tiger was to *both* keep his *Steering Team* abreast of project activities and provide the other teams with the rationale behind the *Steering Team's* decisions.

Some changes in work transfers were made. But, most important, Tiger's job description changed: He was to wear only one hat—manage just his own *Steering Team* group's coordination responsibilities. The three successive DSM spreadsheets had helped guide these decisions.

Because DSM spreadsheets provided the way to visualize and understand their work experiences, they could better meet future schedules, and they could better manage their inevitable organizationally induced iterations and quantify how much information exchange was needed.

- Management saw the affected departmental interfaces.
- By making the dependency structures explicit, they achieved better information flow and division of responsibility.
- When we know what to communicate and to whom, we are more likely to receive what we need when we need it.

Their DSM spreadsheets also

- Identified items likely to be transferred across organizational boundaries.
- Helped management plan how to assign persons with learned critical experience in future projects.

Advanced Engineering	Preliminary Business Case	1																					
	Identify Needs & Objectives		2												BOB								
	Estimate Benefits & Costs			5				X	X			TIGER							BERT				
	Prioritize Alternatives	X		X	6				X														
	Initiate Concept Validation					10				X													
	Design Prototype Hardware						18							X				X	X				
Steering Group	Establish Targets	X	X					3															
	Generate Selection Criteria	X	X					X	4														
	Select Development Direction		X	X	X			X	X	7					TIGER								
	Identify High-risk Failure Modes			X						X	8												
	Allocate Funding & Resources		X			TIGER				X		9											
	Assess Risks Based on Evaluation										X		13	X	X								
Design Audit Team	Conduct Engineering Evaluation	X	X			X		X		X	X	X		11									
	Document Engineering Evaluation			BOB							TIGER			X	12								
	Refine Cost Estimates			X										X		14							
Development Group	Select Application Site												X	X			15						
	Develop Rollout Plan												X	X			X	16					
	Secure Divisional Sign-off																X	X	17				
	Build Prototype Hardware					BERT	X			DAN					BUD					19			
	Conduct Validation Testing																	X			20		
Product Team	Design Production Hardware						X							X							X	21	
	Build Production Hardware																					X	22

FIGURE 9.3
DSM spreadsheet of Figure 9.2 annotated with names of project managers responsible for intergroup coordination.

10

Representing Complexity

Often, we produce models so complicated others are unable to understand or accept them. We also are reluctant to admit that we do not always understand them ourselves.

We need models we all can understand so we can see whether what we have is what we want so we will not waste effort building something we do not really want. The earlier we catch false assumptions, the less effort we waste pursuing them.

A large aircraft manufacturer is constantly reducing cycle times and development costs. The manufacturer is excellent at project management, but it still takes too long to design and develop aircraft. In the 1990s, the manufacturer began a series of projects to reengineer their aircraft design process. The primary goal was to increase productivity by minimizing their design/development cycle time while maintaining highest product quality. Competitiveness in the airline industry requires major improvements in productivity.

Look closely at Figures 10.1 and 10.2; they are different approaches to organize and represent the same engineering process.

In making their case to reengineer their process, managers found that their graph models did not convince senior management that the proposals would address the critical issues. By redrawing their diagrams as DSM spreadsheets, their proposals for change became clear. Figures 10.3 through 10.6 illustrate the differences.

The managers especially wanted to identify decoupled tasks that could be carried out in parallel. They did not know how to recognize them until they made DSM spreadsheets. Although most processes were feed forward, they found many opportunities for parallelism (see Figures 10.5 and 10.6). These reductions came mainly from opportunities for concurrency that were hidden in their graphs.

44 • *Mastering Complexity*

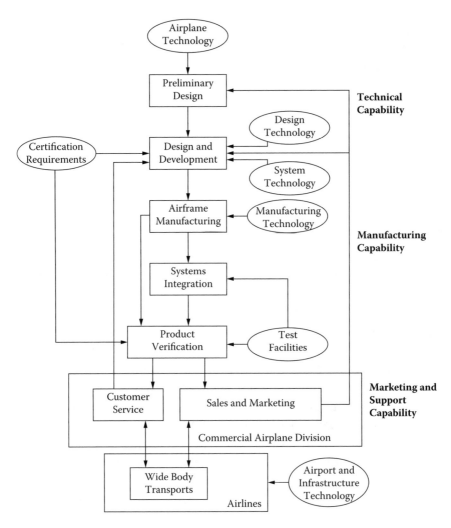

FIGURE 10.1
Original system-oriented design approach diagram.

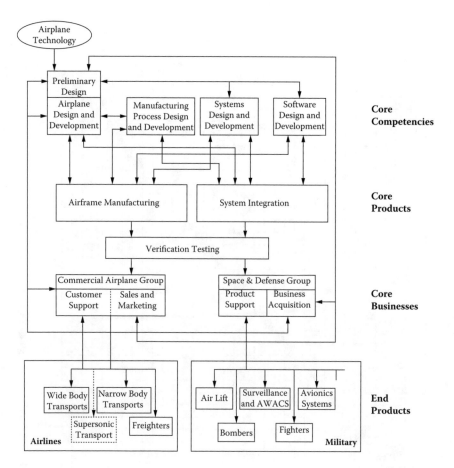

FIGURE 10.2
Original functionally oriented design approach diagram.

46 • *Mastering Complexity*

DISCIPLINE	DESIGN FUNCTION	
Airline customers	Airline company business planning	1
Sales/marketing	Customer liaison	2
Business planning	Business management	3
Research engineering	Airplane technology r&d	4
Airplane project	Preliminary design	5
Business planning	Order entry and scheduling	6
Business planning	Configuration management	7
Design/build team	Detail design	8
Part type centers	Facility planning and maintenance	9
Business planning	Financial planning	10
Manufacturing engineering	Manufacturing capability	11
Manufacturing engineering	Part commonality	12
Manufacturing	Producibility data	13
Manufacturing engineering	QA acceptance test critieria	14
Manufacturing engineering	QA software certification	15
Design/build team	Integrated product definition	16
Manufacturing engineering	Generative process planning	17
Part type centers	Manufacturing resource planning	18
Part type centers	Material handling	19
Manufacturing engineering	Material	20
Manufacturing engineering	Nc program generation	21
Part type centers	Numerical control devices	22
Vendors	Outside resource part type	23
Part type centers	Quality measurements	24
Part type centers	Tool and equipment inventory	25
Part type centers	Unscheduled maintenanece	26
Manufacturing engineering	Quality assurance history	27
Certification and testing	Laboratory testing	28
Certification and testing	Certification testing	29
Airline customers	Flight operations liaison	30
FAA	Government/industry liaison	31
Customer support	Maintenance support	32
Customer support	Problem reporting	33
Certification and testing	Acceptance testing	34
Finance	Accounting	35

FIGURE 10.3
DSM spreadsheet showing task sequence for original functionally oriented design approach.

FIGURE 10.4
Design schedule for functionally oriented process before reengineering program.

48 • *Mastering Complexity*

DISCIPLINE	DESIGN FUNCTION	
Airline customers	Airline company business planning	1
Research engineering	Airplane technology R&D	2
Part type centers	Facility planning and maintenance	3
Research engineering	Manufacturing technology R&D	4
Research engineering	Software technology R&D	5
Research engineering	Systems technology R&D	6
Sales/marketing	Customer liaison	7
Business acquisition	Order entry and scheduling "I"	8
Airplane project	Preliminary design "I"	9
Business acquisition	Business management	10
Business acquisition	Configuration management	11
Manufacturing engineering	Manufacturing capability	12
Manufacturing engineering	Part commonality	13
Manufacturing engineering	QA acceptance test criteria	14
Manufacturing engineering	QA software certification	15
Business acquisition	Financial planning	16
Manufacturing	Producibility data	17
Design/build team	Detail design	18
Design/build team	Integrated product definition	19
Manufacturing engineering	Generative process planning	20
FAA	Government/industry liaison	21
Manufacturing engineering	NC program generation	22
Part type centers	Tool and equipment inventory	23
Part type centers	Manufacturing resource planning	24
Part type centers	Material handling	25
Manufacturing engineering	Material	26
Part type centers	Numerical control devices	27
Vendors	Outside resource part type	28
Part type centers	Quality measurements	29
Manufacturing engineering	Quality assurance history	30
Certification and testing	Laboratory testing	31
Certification and testing	Certification testing	32
Airline customers	Flight operations liaison	33
Certification and testing	Acceptance testing	34
Customer support	Maintenance support	35
Finance	Accounting	36

FIGURE 10.5
DSM spreadsheet showing task sequence for reengineered process. (Note the opportunities for previously unseen parallelism shown by the bars.)

Representing Complexity • 49

DISCIPLINE	DESIGN FUNCTION
Airline customers	Airline company business planning
Research engineering	Airplane technology R&D
Part type centers	Facility planning and maintenance
Research engineering	Manufacturing technology R&D
Research engineering	Software technology R&D
Research engineering	Systems technology R&D
Sales/marketing	Customer liaison
Business acquisition	Order entry and scheduling "I"
Airplane project	Preliminary design "I"
Business acquisition	Business management
Business acquisition	Configuration management
Manufacturing engineering	Manufacturing capability
Manufacturing engineering	Part commonality
Manufacturing engineering	QA acceptance test criteria
Manufacturing engineering	QA software certification
Business acquisition	Financial planning
Manufacturing	Producibility data
Design/build team	Detail design
Design/build team	Integrated product definition
Manufacturing engineering	Generative process planning
FAA	Government/industry liaison
Manufacturing engineering	NC program generation
Part type centers	Tool and equipment inventory
Part type centers	Manufacturing resource planning
Part type centers	Material handling
Manufacturing engineering	Material
Part type centers	Numerical control devices
Vendors	Outside resource part type
Part type centers	Quality measurements
Manufacturing engineering	Quality assurance history
Certification and testing	Laboratory testing
Certification and testing	Certification testing
Airline customers	Flight operations liaison
Certification and testing	Acceptance testing
Customer support	Maintenance support
Finance	Accounting

FIGURE 10.6
Design schedule for reengineered process.

Section 3

Building Dependency Maps

"Begin at the beginning," the King said very gravely, "and go on till you come to the end—then stop."

—Lewis Carroll

11

What Depends on What?

11.1 INTRODUCTION

To begin to understand what depends on what, put a list of elements together. Their initial order is unimportant and creating an initial DSM spreadsheet can be a *top-down* or a *bottom-up* process.

In a DSM spreadsheet, each element will be represented—once as a row and again as a column. Elements are listed down the left-hand side of the DSM spreadsheet. The same list of elements is placed along the top of the DSM spreadsheet, from left to right. Their row order and their column order are identical. Each element's row and column intersect at the DSM spreadsheet diagonal cell.

- **Needs give marks in rows, and feeds give marks in columns.**

Where an element's results are used by another element gives the marks in the columns of the DSM spreadsheet.

Where an element needs an output from another gives the marks in the rows of the DSM spreadsheet.

In Figure 11.1, the mark in row **A**, column **D**, means **A** depends on **D**. If no dependency relationship exists, we do not put a mark in that row. For example, **D** has no dependencies.

FIGURE 11.1
DSM with dependencies shown.

11.2 SOME DEPENDENCY TYPES

- TASK DEPENDENCY

A mark in the DSM spreadsheet shows where one task depends on another.

- UNCERTAINTY DEPENDENCY (I particularly use this kind)

We are unsure if **A** is dependent on **B**. We do not want to ignore risks caused by our uncertainty. So, initially we put a mark to indicate possible dependency. Only after analysis do we keep or remove the mark.

- INFORMATION DEPENDENCY

Information required by **B** is produced by **A**.

- RESOURCE DEPENDENCY

Any necessary resource is also a dependency. Someone might have a particular skill. Other resources might include materials or required tools.

- PREFERENTIAL DEPENDENCY (see Chapter 10 example)

This dependency is the result of preferences or policies of an organization.

- FUNCTIONAL DEPENDENCY

A functional requirement in an organization exists between **A** and **B**.

- MIXED DEPENDENCIES

And last, dependencies can be anything. A DSM need not have all its elements of one type.

One DSM spreadsheet we used for a new consumer product simultaneously contained such diverse dependent elements as styling and ergonomics guidelines, company executives' views of the market environment, and corporate accounting and capital equipment budgeting rules.

- **How many dependencies are necessary to include?**

Include only direct dependency pairings.
Indirect dependencies emerge automatically.

11.3 DIRECT DEPENDENCIES

- What each element needs and the source of what each element needs is shown by marks in row cells.
- Where the result of a column element is used by the row element is shown by marks in column cells.
- A mark in a cell below the diagonal means a later element uses something from a prior element.
- A mark in a cell above the diagonal means something from a later element is required to *complete* a prior element.
- If an element is dependent on other elements, it must be performed sequentially.
- If an element is independent of another element, they could be performed concurrently (in parallel).

After all the elements are listed across the top and down the left-hand side of a DSM spreadsheet, we proceed row by row.

- From left to right, we move along the row and ask, "Does this row need anything from the element in this column?" If it does, we put a mark in the corresponding DSM spreadsheet cell.

- What about the level of detail? Start lean.

We try to include only necessary and sufficient dependencies. We initially avoid weak or unnecessary dependencies because they can make our DSM spreadsheet harder to understand. Ask, "Is this necessary to be in our plan?"

One way to do this is to construct the DSM spreadsheet *backward,* from finish to start. Going backward means it is easier to challenge our assumptions as we go. Often, dependencies are put in place because, "We've always done it that way." The more we can challenge and eliminate, the more we are likely to include only those dependencies that are important.

- It could seem to be an unlimited number of things that seem reasonable to include. No, they should connect to our objective.

The resulting DSM spreadsheet will probably look something like a shotgun blast, with marks above and below the diagonal. Again, see the array of dependencies in Figure 11.1.

Do not worry whether the initial order of elements in our DSM spreadsheet is in a logical order. We have software to rearrange them. (See Part 6 of this book for software tools to use.)

We do a first-pass check for completeness by going across rows and down columns to inspect for presence or absence of marks.

- Often, we do not need to do anything more.
- Reading across a DSM spreadsheet row, we check if we have captured all the elements linked to the row's requirement.
- For each element, scan down its spreadsheet column. This displays all its *dependent elements.* A mark in the row element means this is a *dependent element.*

Working with DSM spreadsheets in this way can point to where hard numbers are required and where they are not.

- **It is the qualitative nature of these first logical steps, requiring minimal quantitative data, that potentially widens DSM spreadsheet applications beyond many methods that require hard numbers for understanding.**

- Once we have completed our initial DSM spreadsheet, we can check it for consistency:
  "Is this reasonable?"
  "No predecessors?"
  "No successors?"

Before, we might have known what we needed. We might not have been entirely sure who needs our work—who needs what we generate. Now, with our DSM spreadsheet, regardless of the order of elements, we can know who needs our work.

- Our DSM spreadsheet already shows us something about the communication that must occur. Once we build up the DSM spreadsheet by rows, we then have the columns. The columns tell us where something is needed. DSM spreadsheets identify each element and every other element affecting it in some way. DSM spreadsheets show where elements are affected by any change in any other element.

Often, benefits are realized just by participating in DSM spreadsheet map construction.

After hearing my DSM presentation, one client spent an entire day attempting, on their own, to build a DSM spreadsheet.

- **They did not succeed in actually creating any DSM spreadsheet that day.**
- **But, more important, they left for home satisfied—for the first time they clearly understood the problem they had been facing for a long time, previously without success.**

By making dependency structures explicit and understandable, we can achieve better information flow and division of responsibility. We can better understand and manage our domain and see clearly our interfaces with others. We know what to communicate, when, and to whom. We can expect to receive what we need when we need it.

These questions might have to be asked repeatedly. "What do I need?" has to be asked every time there is a major change. If we ask this question, we will soon come to understand both what we need and what we owe. Then, we can organize both.

Let us move on and see how DSM spreadsheet maps help us organize and manage some business processes.

12

Projects and Structure

Suppose our DSM spreadsheet represents a set of tasks. The order of elements is the task sequence. Marks in a row of the DSM spreadsheet represent all of the tasks whose output is required to perform the task corresponding to that row. Reading down a specific column reveals which tasks receive something from the task corresponding to that column. Marks below the diagonal represent a forward transfer to later (i.e., *feed-forward*) tasks. Marks above the diagonal depict feedbacks to earlier listed tasks (i.e., *feed-back* tasks) and indicate that an upstream task is dependent on a downstream task as shown in Figure 12.1.

- An essential difference between traditional workflow tools—unlike PERT (program evaluation and review technique) and Gantt charts—DSM spreadsheets allow us to treat the relationship between task elements in any order, not just sequentially. No matter what order a task is in, its dependencies remain the same.

Let us illustrate. A manufacturer of microfilm imaging equipment approached a film company to design and supply microfilm cartridges for the manufacturer's new machine.* Specifications were similar to products developed by the film company's cartridge group. Their usual development time was 26 months, but the customer needed prototype cartridges for a trade show in just 8 months. The film company accepted and met the challenge of cutting its normal development time. Use of a DSM spreadsheet for project management was crucial to its success. We show a shortened version of the complete project plan's three key components—workflow, logic, and schedule—in Figures 12.2, 12.3, and 12.4, respectively.

---

* Adapted from Ulrich, Karl, and Eppinger, Steven. 2003. *Product Design and Development.* New York: McGraw-Hill/Irwin.

60 • *Mastering Complexity*

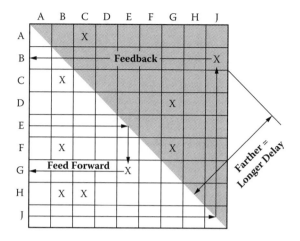

FIGURE 12.1
Basic DSM spreadsheet dependencies.

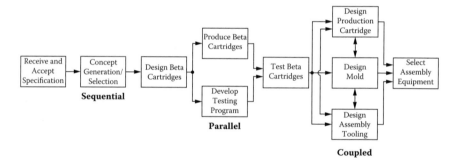

FIGURE 12.2
Workflow.

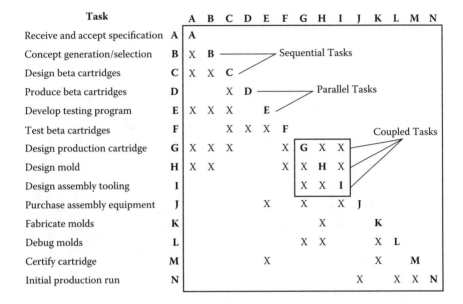

FIGURE 12.3
Logic.

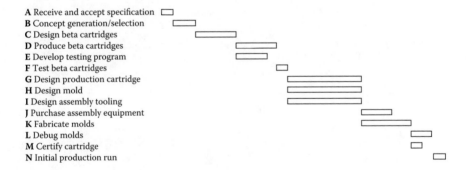

FIGURE 12.4
Schedule.

13
Making Assumptions

To obtain a workflow, we need to choose how to order the DSM spreadsheet elements. Marks below the DSM spreadsheet diagonal correspond to the task diagram showing where a task gets what it needs from an earlier task. Marks above the DSM spreadsheet diagonal correspond to where a task must use an assumption because the task that produces the information it depends on cannot be done until later. Marks below the diagonal could form a *critical path* for a project schedule if all our assumptions prove to be correct. But, this schedule might need to be modified by adding reviews should some assumptions prove wrong.

Each reordering of rows and their corresponding columns in DSM spreadsheets corresponds to a different approach. We could schedule project tasks based on this approach. Thus, without using a DSM spreadsheet for our project, we could have completed tasks in the order of the DSM spreadsheet in Figure 13.1, completing tasks **A**, **B**, and **C** only to arrive at task **D** and realize that we have made some mistake, forcing us to begin again.

A DSM spreadsheet helps us to cut out unnecessary rework and wasteful iteration. We can organize work so that the information needed to begin a task would be ready just in time. The reordered DSM spreadsheet in Figure 13.2 shows better task arrangement.

We rearranged the DSM spreadsheet to move as many elements as possible from above the diagonal to below the diagonal. This reordering often does all that needs to be done. A rearrangement of the elements alone removed all the unnecessary feedbacks. In Part 4 of this book, we explain in some detail how to do this.

Only one mark is above the diagonal. What do we do about this mark? If this DSM spreadsheet represented a task sequence, to make progress at step **F**, we would need information about something that has not happened

64 • *Mastering Complexity*

	A			X		
X	**B**	X				
X		**C**				X
			D			
		X			**E**	X
				X	X	**F**

FIGURE 13.1
Original DSM spreadsheet with three apparent feedbacks.

yet. If we want to schedule the dependency sequence **D, A, F, C, E, B**, we must assume an initial value for **F**. Again, see Figure 13.2.

- The mark above the DSM spreadsheet diagonal shows we require an assumption to schedule our project sequentially. Marks below the DSM spreadsheet diagonal all have sequentially satisfied dependencies.
- With this DSM spreadsheet, we see what needs to be our focus. Even for a well-understood process, there are potential areas of risk. In new programs, because our process might be less understood, the risk is much higher.

For any particular problem, the dependency structure of the problem will remain the same, while the approach can easily be changed. This gives us the opportunity to play with various approaches with their different task orders and assumptions.

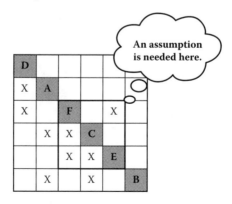

FIGURE 13.2
Reordered DSM spreadsheet with only one necessary feedback.

In further sections of the book, we expand on these ideas and provide more examples.

14

Manipulating Assumptions

In projects, we can account for different levels of risk using strengths of dependency. Once we classify dependencies within a DSM spreadsheet, we can use them to reorder DSM spreadsheet elements or assess risks associated with our schedule approach.

Each DSM spreadsheet element is assigned a risk level of **A** for *strong*, **B** for *moderate*, and **C** for *weak*. This is summarized in Figure 14.1.

- Often, any weak dependencies can be safely omitted entirely from our DSM spreadsheet.

To show one use of dependency classification, we look at building construction. Building construction is perhaps the most mature of all engineering disciplines. If any area is a fully rational and optimized endeavor, it should be building construction.

The as-is design process of a commercial building project was modeled using DSM spreadsheets shown in Figure 14.2.* However, in practice, building designers release information to the timing required by contractors. The usual practice is to plan the design process backward from the date when its deliverables are due to be released to the client or contractor.

- This may affect the quality of design information available during construction.

---

* Adapted from Austin, Simon, Baldwin, Andrew, Li, Baizhan, and Waskett, Paul. 2000. Application of the Analytical Design Planning Technique to Construction Project Management. *Project Management Journal*. June: 48–59.

68 • *Mastering Complexity*

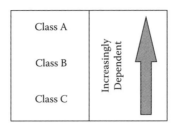

FIGURE 14.1
Dependency risk classifications.

So, even though a better schedule might be produced using DSM spreadsheets, building design is often programmed to suit construction. Any problems that result are eventually resolved, at high cost, on site.

Suppose we choose to move properly sequenced high-risk tasks? Figure 14.3 shows a DSM spreadsheet in which two **A** risk-level tasks have been moved up in the schedule so *materials procurement* for the building foundation pouring can be *fast tracked*. This provided the alternative, more efficient, *could-be* sequencing of design tasks shown in Figure 14.3. This is often common practice. But, there is increased risk (e.g., safety); the foundation will not meet requirements unless we spend more money to overbuild to compensate for our lack of a robust design.

- Special consideration must be given to the design information requirements.
- We have to estimate a suitable value because it cannot be revised during construction. For example, choosing to pour a much stronger foundation footing than design might later suggest is required.

In a large commercial building, there can be as many as 500 identifiable major tasks and 2600 to 3000 dependencies in the detailed stage of building design.

Any proposed changes to the optimal design program can be reviewed to establish the ease with which task duration and resources can be reallocated and the most suitable pieces of information to estimate. Also, the additional cost incurred through overdesigning some elements of the building can be compared to the costs of extending the duration of the corresponding work packages.

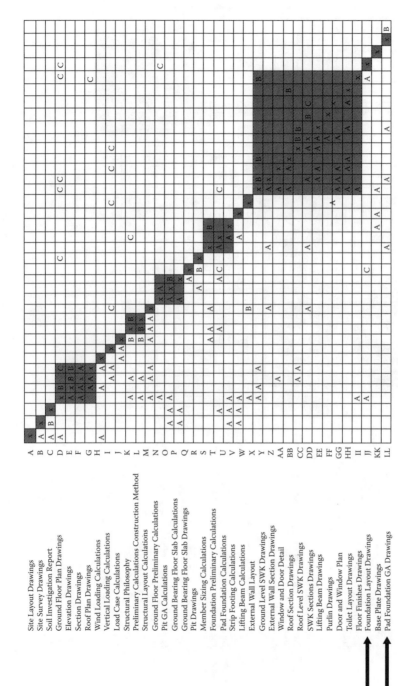

FIGURE 14.2
Classified and partitioned construction information requirements.

70 • *Mastering Complexity*

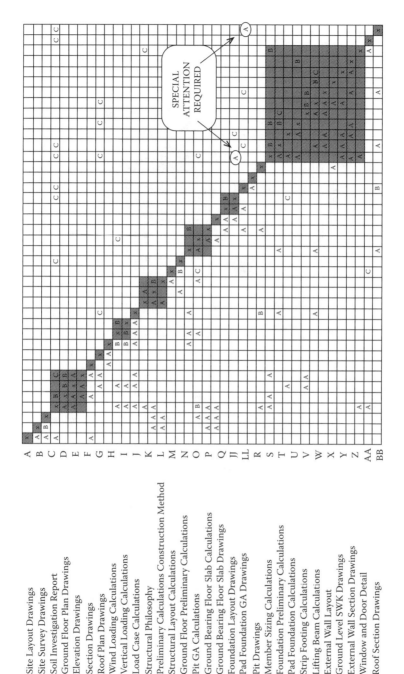

FIGURE 14.3
Possible fast-tracking construction issues.

15

Active Risk Management

Businesses often take risks to bring their products to market faster, pushing the technology envelop hard and shortening design cycles. The spoils go to those who successfully manage these risks.

When we develop new products, there are always things we need now that will not be known until we are further along in the process. But, assumptions can be wrong, delaying product introduction and causing embarrassment as the expectations of suppliers and their customers are dashed.

We might have no idea in advance how much iteration might be required until we obtain an adequate result. Iteration introduces schedule and budget risks.

- A DSM spreadsheet helps us address the riskier elements first. To minimize these risks, a DSM spreadsheet helps us to decide where we will need assumptions and where iteration might be needed.
- By recognizing and managing our assumptions, we can manage projects from where they are rather than from where they were expected to be.
- When we plan, a DSM spreadsheet not only shows what assumptions we make but also when in the plan they are ready to be reviewed. Just how many loops of iteration remains uncertain, but by knowing which assumptions have been resolved and which have not, risk is reduced.

Because our assumptions are made explicit, we can control them better—see where they need to be verified, recognize how our plan must change if they prove to be wrong, and know at all times what still depends on (as-yet) unverified assumptions.

We will not overlook some assumption, throwing our plan, in both time and cost, completely out of kilter.

- Iterations might still occur when an assumption later proves to have been wrong.

If any assumptions proved later to be different from our earlier choice, it might be necessary to go back to a previous task that used this assumption and redo it, using the revised assumption.

The process stops when the difference between required values is close to the value from the previous iteration.

A wholesale distributor decided to replace its original mainframe warehouse inventory management system with a modern, lower-cost UNIX system.[*] The company had some unique requirements. The hardware vendor, the software vendor, and the value-added reseller all claimed that adding customized features to an existing software package would meet the distributor's key performance requirement: to be able to batch process up to 10,000 new orders overnight. The initial project plan for the 2-year project used a waterfall development model. Figure 15.1 shows the schedule for the project phases. A waterfall development process works well if there is a *stable* set of requirements and well-understood technologies. Projects do not start work on the next phase until all the questions in the current phase are resolved and no new information will require the project to redo work in more than the immediately preceding phase.

In this case, design problems surfaced after 23 months into the project when the initial system test of the warehouse database batch-update process took over 26 hours to complete.

Phase	Duration
Requirements Analysis	3 Months
High-Level Design	1 Month
Detailed Design	4 Months
Coding, Debugging, and Data Conversion	15 Months
System Testing	1 Month

FIGURE 15.1
Initial project plan.

---

[*] This example is adapted from an unpublished work by Hugh McLaughlin.

The project was ultimately cancelled, but not before valiant efforts to rescue it, including the following:

- The hardware vendor providing, at its expense, a computer with twice the capacity and speed, and
- Senior developers reworking database tables, indices, and code for an extra 6 months.

Risks are always crystal clear in hindsight. But, the moral of the story is that four major parties—the customer, hardware vendor, software vendor, and value-added reseller—all accepted too large a risk by allowing the initial system test to be scheduled only after the design was complete. What was missing was better understanding of the assumptions that had been made and when and how they should be verified.

The DSM spreadsheet in Figure 15.2 explicitly identifies the information dependencies and suggests an alternative project phase sequence. By starting project planning with information dependencies rather than with the more usual approach of workflow dependencies, they could have made a better trade-off between assumptions and tasks. For less than an 8-percent increase in the original budget and no stretching out the initial completion date, it would have been possible to test the nightly batch-processing requirements at the end of month 4.

Phases	Months	A	B	C	D	E	F	G	H	I	J	K	L
Requirements Specification	1.00	A	X		X								
Data Map Design	1.00	X	B										
Database Tables Creation	0.50		X	C									
Batch-Update Test	1.50	X		X	D								
Base-System Test	0.10	X		X	X	E							
Test Analysis	0.30	X				X	F			X			
Query and Tables Redesign	1.00	X					X	G		X	X		
Data Tables Reload	0.20			X			X		H				
Batch-Update Retest	0.10						X	X	X	I	X		
Full-System Test	0.10	X			X				X	X	J		
Requirements Confirmation	2.00	X	X				X				X	K	
Final Acceptance	1.00	X	X				X				X	X	L

FIGURE 15.2
Revised project plan DSM spreadsheet.

In Figure 15.2, the first block representing phases **A** through **E** defines the first development cycle. The block **F** through **J** represents full-system testing and any necessary design iterations. If the critical, nightly batch-update processing requirement is met with ease, then there is no need to iterate phases **F** through **J**.

The information feedback from phase **J** to phase **F** is especially critical: The total night batch-update time requirement must be met. Initially, phase **F** accepts the results of phase **E**. On any subsequent iteration to reduce total batch-update process time, phase **F** would require information from phase **J** to enable the project to choose where to focus. This is likely to vary from iteration to iteration of the analysis–redesign–test iteration loop (phases **F** through **J**) depending on performance improvements made by previous iterations.

By resequencing and highlighting the potential feedback loops associated with meeting performance requirements, they could have reduced the risk of show-stopping failures late in the project. It would have been possible to determine whether the key system batch-update completion requirement was achievable and, if not, abort the project as early as 6 months after the start.

16

Unwrapping Circuits

Standard project scheduling techniques make no explicit provision for handling dependency circuits—circular loops of tasks. So, how do we design project plans that we can schedule? A DSM spreadsheet can show us the way. By making assumptions that are both sensible and break loop structures, we can unwrap the elements in a circuit into sequences that can be scheduled and iterated as necessary.

Consider the design of an electric car. Element dependencies describe the car and how it is made. Together, these relationships form a *descriptive statement* of the design constraints. To design the car, our tasks will set values on these elements. The project to do this is a *prescriptive process*.

Most projects' schedules will be constrained by the underlying project information dependencies and their logical structure.

- In Figures 16.1 through 16.7, *lowercase letters* in the DSM spreadsheet denote information; *uppercase letters* denote tasks that use or produce the information.

Figure 16.1 is a graph illustrating the information dependencies in the design of an electric car. (Ignore the vertical bar between **i** and **e** for a moment.) Figure 16.1 indicates that

- **BATTERY WEIGHT g** depends on STORED ENERGY **e**.
- **AERODYNAMICS b** depends on BATTERY WEIGHT **g**.
- **TOTAL WEIGHT d** depends on BATTERY WEIGHT **g** and AERODYNAMICS **b**.
- **ACCELERATION i** depends on AERODYNAMICS **b** and Total Weight **d**.
- **STORED ENERGY e** depends on ACCELERATION **i**.

76 • *Mastering Complexity*

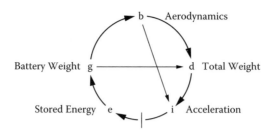

FIGURE 16.1
Constraints between design elements.

The information graph of Figure 16.1 can be represented as a *descriptive* DSM spreadsheet in Figure 16.2.

- But, which design element should we start with?

Having created the *descriptive* DSM spreadsheet of Figure 16.2, we can use an estimate for STORED ENERGY e_1 when we do not know a value for ACCELERATION **i**. E_1 represents the design task **E** that uses the first estimate of **e** (e_1) made without having a value for **i**. This is equivalent to breaking apart the arc between (we call this *tearing*) the graph in Figure 16.1 at the point of the vertical bar between **e** and **i**. (How we decided to tear between **i** and **e** is covered at length in Part 4.)

A reordering gives the *prescriptive* task DSM spreadsheet shown in Figure 16.3. **E, G, B, D,** and **I** represent design tasks required to determine elements **e, g, b, d,** and **i**. Figure 16.4 *prescribes* the possible project task sequence (E_1, **G, B, D,** and **I**) to design elements (**b, d, i, e,** and **g**) described by the DSM spreadsheet in Figure 16.3.

By tearing, we unwrap the dependency circuit into a sequence of task iterations as shown in Figure 16.4.

Aerodynamics	**b**				X
Total Weight	X	**d**			X
Acceleration	X	X	**i**		
Stored Energy			X	**e**	
Battery Weight				X	**g**

FIGURE 16.2
DSM spreadsheet of design element dependencies.

	Stored Energy	E				E_1
	Battery Weight	X	G			
	Aerodynamics		X	B		
	Total Weight		X	X	D	
	Acceleration			X	X	I

FIGURE 16.3
DSM spreadsheet of reordered design task dependencies.

$$E_1(e_1) > G_1 > B_1 > D_1 > I_1 > \text{Review}_1 > E_2 > G_2 > B_2 > D_2 > I_2 > \text{Review}_2$$

|_____Iteration 1_____| |_____Iteration 2_____|

FIGURE 16.4
One plan for design tasks.

- The resulting project plan does not contain circuits. Now, we can use it to develop a schedule for the tasks.
- We also know exactly what estimates we used and where we used them to accomplish this.

After each iteration, we insert reviews to check whether the estimate for element **e** was adequate or whether another iteration is required. In Figure 16.4, we began with task E_1 by using an estimate e_1. Our first review, **REVIEW$_1$**, assesses whether this estimate was satisfactory. If not, we make another iteration.

Suppose we were also to decide to estimate AERODYNAMICS (b_1). e_1 and b_1 are the first-pass estimated values for design elements **e** and **b**, respectively. We use estimates b_1 and e_1 for tasks B_1 (b_1) and E_1 (e_1) shown in Figure 16.5.

To possibly accelerate our project, we might make an additional tear between elements **g** and **b** to show a way to gain more parallelism. This gives the plan for design tasks shown in Figure 16.6.

But, if after the first review we find that the value for element **b** must be revised from its original estimated value (b_1), we could revert to the sequence shown in Figure 16.7.

Stored Energy	**E**				**E$_1$**
Battery Weight	X	**G**			
Aerodynamics		**B$_1$**	**B**		
Total Weight		X	X	**D**	
Acceleration			X	X	**I**

FIGURE 16.5
Alternate DSM spreadsheet showing design tasks.

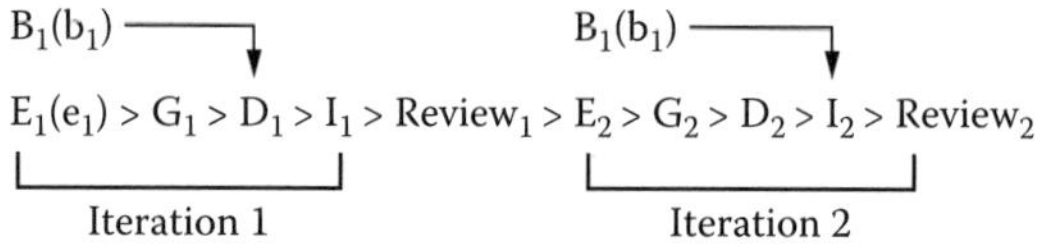

$$B_1(b_1) \longrightarrow \quad\quad B_1(b_1) \longrightarrow$$
$$E_1(e_1) > G_1 > D_1 > I_1 > \text{Review}_1 > E_2 > G_2 > D_2 > I_2 > \text{Review}_2$$
$$\underbrace{\phantom{E_1(e_1) > G_1 > D_1 > I_1 > \text{Review}_1}}_{\text{Iteration 1}} \quad \underbrace{\phantom{E_2 > G_2 > D_2 > I_2 > \text{Review}_2}}_{\text{Iteration 2}}$$

FIGURE 16.6
Another plan for design tasks.

$$B_1(b_1) \longrightarrow$$
$$E_1(e_1) > G_1 > D_1 > I_1 > \text{Review}_1 > E_2 > G_2 > B_2 > D_2 > I_2 > \text{Review}_2$$

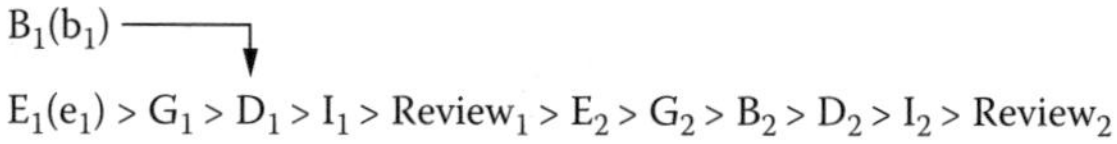

FIGURE 16.7
Yet another plan for design tasks.

Section 4

Exposing Logical Structure

God give us the grace to accept with serenity the things that cannot be changed, courage to change the things which should be changed, and the wisdom to distinguish the one from the other.

—**Reinhold Niebuhr**

17
Topological Order

Topology, a branch of geometry, deals with properties of connected graphs. Topology describes how the physical arrangement of connections also links nodes logically—how they depend on each other.

- Whenever an element is encountered, all its predecessor elements on which it depends have already been encountered earlier, we have *topological order.*

Topological order implies the situation or process being modeled is logically sequential—its *causes precede their effects.* No graph arcs go backward from later elements to earlier elements (see Figure 17.1).

- Are most business processes intrinsically sequential? Generally, this is not the case.
- Once we realize any process can be represented as a square DSM spreadsheet, we can show how its graph topology and its matrix properties could prevent some desired outcomes.

Clearly, topological order does not exist if dependency relationships have any closed paths (*circuits*). This means both **A** precedes **B** and **B** precedes **A**. Topologically, no dependency relationship between **A** and **B** is OK. **A** before **B** is OK. **B** before **A** is OK. But, **A** before **B** and **B** before **A** are not OK. These topological constraints are illustrated in Figure 17.2.

Elements making up closed paths can be grouped into blocks in a DSM spreadsheet. They are called *blocks* because they appear as square arrays of marks symmetrically astride the DSM spreadsheet diagonal. A block is an indivisible tangle of mutually dependent elements. If we were to pick any

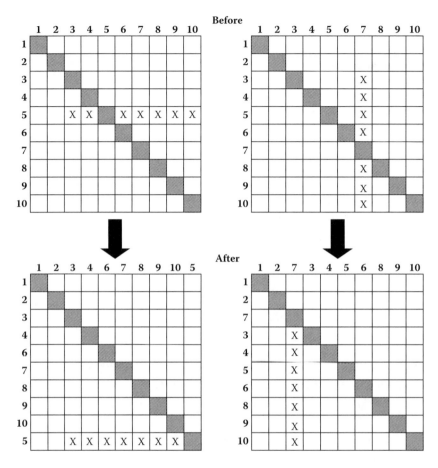

FIGURE 17.1
Two simple topological reorderings.

element in a block, we can get from that element to every other element in the block and back again (i.e., a circuit).

- Identifying true circuits can be difficult, if not impossible, using a graph diagram but can be simplified considerably using a DSM spreadsheet.
- We can know where the true circuits are—not those we have artificially created.
- Graphs might only appear to contain circuits.
- If a graph contains any true circuits, it is not possible to arrange all its nodes in a topological order without some intervention.

FIGURE 17.2
Topological reality constraints.

Where true circuits do occur, we have tools and techniques to deal with them. *Tearing* is what we call the method by which we address situations that initially violate topological order. That is the topic of Chapter 19. We present an overview of the techniques there. In Part 6, we describe software that can help us accomplish this task more easily.

18
Partitioning

Suppose we have a long cord that is kinked and apparently knotted. We can, after some work, remove all the apparent knots to have an unkinked cord. If there is also a tightly knotted clump somewhere along the cord, we can still unravel the entire cord. This clump is in *its proper topological place* along the cord. **But, we cannot unwrap it.**

We call this unraveling process of topological reordering DSM spreadsheet rows and columns *partitioning*. In any reordering of a DSM spreadsheet, we move both a row and its corresponding column together the same amount in the same way. Partitioning identifies the elements in circuits and clusters them into blocks. DSM spreadsheet reordering might not completely eliminate circuits, but it minimizes them and moves them as close as possible to the DSM spreadsheet diagonal.

Except for the order of elements within blocks, we reorder the entire DSM spreadsheet. All individual elements and blocks are put into topological order. In the final DSM spreadsheet, every element and block is preceded only by those elements encountered earlier. No arcs go from a later element or block to an earlier element or block.

If nodes are on parallel paths, a DSM spreadsheet says nothing about the relative order of these paths. So, the order of some elements as they appear in a DSM spreadsheet is arbitrary.

Once partitioning is completed, tearing can be used to order elements within the blocks. Together, partitioning and tearing tools can be used to restructure our business process to facilitate the topological flow. Figure 18.1 shows the result of partitioning.

Free downloadable spreadsheet add-ons and commercial software tools to accomplish the tasks of partitioning and rearranging rows and columns are described in Part 6 of this book.

86 • *Mastering Complexity*

	Task A	Task B	Task C	Task D	Task E	Task F	Task G	Task H	Task I	Task J	Task K	Task L	Task M	Task N	Task O	Task P	Task Q	Task R	Task S	Task T
Task A	A																			
Task B	X	B																		
Task C	X	X	C																	
Task D	X		X	D																
Task E		X			E								X							
Task F		X	X			F											X		X	
Task G		X		X			G													
Task H	X				X			H												
Task I			X	X		X			I											
Task J		X	X						X	J										
Task K								X			K			X						
Task L											X	L							X	
Task M								X					M	X					X	
Task N			X								X			N	X					
Task O		X									X				O					
Task P														X	X	P				
Task Q															X		Q	X		
Task R																	X	R		
Task S							X												S	
Task T													X							T

Arbitrary Order

	Task A	Task B	Task C	Task D	Task G	Task J	Task I	Task S	Task L	Task E	Task O	Task M	Task Q	Task R	Task F	Task H	Task K	Task N	Task P	Task T
Task A	A																			
Task B	X	B																		
Task C	X	X	C																	
Task D	X		X	D																
Task G		X		X	G															
Task J		X	X			J	X													
Task I			X				I			X										
Task S								S	X											
Task L						X		X	L											
Task E		X								E										
Task O		X									O									
Task M							X	X				M								
Task Q											X		Q	X						
Task R													X	R						
Task F	X	X													F				X	
Task H	X														X	H				
Task K															X	X	K	X		
Task N				X													X	N	X	
Task P										X		X						X	P	
Task T												X								T

Partitioned

FIGURE 18.1
An example of topological partitioning.

19
Tearing Circuits

19.1 INTRODUCTION

When partitioning identifies blocks, each can be subjected to a second level of analysis we call *tearing*. We use tearing to topologically order blocks internally.

We tear circuits because they violate topological order. Tearing orders marks in a DSM spreadsheet so when the DSM spreadsheet is repartitioned, only general marks (**x**) remain below the diagonal, as shown in Figure 19.1.

Although a DSM spreadsheet is more compact and easier to work with, to understand tearing we choose to describe it in terms of its equivalent graph. In this graph, a directed arc between nodes is represented within DSM spreadsheets by elements having predecessors' marks in their rows. Removing an arc in the graph corresponds to deleting the mark in the cell at the corresponding row and column intersection in DSM spreadsheets. Graph arcs or equivalent DSM spreadsheet cell marks that we remove are called *tears*. A torn arc means we have broken a connection to that element.

We can either tear arcs ending with a node or tear arcs beginning with a node. Tearing arcs going into a node removes the dependence on predecessors to this node. Tearing arcs leaving a node removes it as a predecessor to other nodes.

How do we select arcs for tearing? For complex situations, especially those for which we have little prior understanding, we often start with a purely structural approach. In this way, we can systematically tear blocks in our DSM spreadsheet into smaller parts. Once we have achieved some level of decomposition of our DSM spreadsheet, we are better able to know what we have and what we would choose to do next.

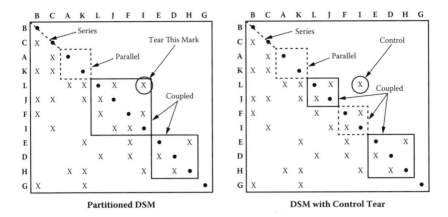

FIGURE 19.1
Tearing marks in a DSM spreadsheet.

- Letting software suggest structural choices can be an extremely productive exercise. It can open our eyes—pointing the way to new and possibly more effective problem solutions.

There are no rigorous criteria to help us choose the best tears. So, we often generate several tears before settling on those that look good. For each reordering of rows and columns in a DSM spreadsheet, the element-to-element dependencies remain unchanged as their order changes. This gives the opportunity to safely try reordering. Examine each result to learn.

We use software tools to systematically find and tear circuits and topologically reorder DSM spreadsheets. Part 6 of this book describes several software tools that could be used.*

- After we choose to tear elements, we should also consider which tearing makes sense in terms of our situational knowledge. The marks we choose to tear can represent assumptions, critical control, or feedback dependences.

Should we tear elements in a row? Should we tear elements in a column? Some rows might represent elements where key estimates or assumptions are made. We could choose to tear in these rows. We could tear columns

---

* Don Steward's own software tool PSM 32 is described at length in Chapter 26.

to reduce the number of different nodes whose results must be assumed. Or, we might do only a small number of tears altogether.

19.2 WHAT TEARING DOES

- Where an element's results are used by another element gives marks in columns. If we tear elements in a column, we weaken influence on its successors. The next time our DSM spreadsheet is partitioned, the row corresponding to the torn column moves down in the block.
- Where an element needs results from another element gives marks in rows. If we tear elements in a row, we weaken influence of its predecessors. The next time our DSM spreadsheet is partitioned, the torn row moves up in the block.

19.3 PROBLEM-SOLVING ADVICE

- Analyze blocks until the torn circuits represent practical sequences of elements satisfying both topological ordering and our situational knowledge. Keep in mind what it means to use assumptions for some elements. They are used to determine other elements.
- Once you understand our original problem better, you might see where you should have selected elements or dependencies differently. Make these changes and start the topological analysis again.
- Show your DSM spreadsheets to others to obtain their ideas and if changes need to be made. "Does this DSM spreadsheet and the solution it suggests make sense?"

20

Structural Components

- PRINCIPAL CIRCUITS

A *Principal Circuit* is the longest circular path through a block. As the largest circuit of dependent nodes in a block, a Principal Circuit defines the size of the block. Principal Circuits are the most important circuits in our DSM spreadsheet. Until we identify, characterize, and tear these circuits, we might not be finished.

- SHUNTS

Sometimes blocks do not have just one set of circular dependencies; they have several. *Shunts* cause these other dependencies.

After the original Principal Circuit is torn, some shunts remaining might still form shorter, inside circuits as in the graph shown in Figure 20.1. Shunts are paths that do not include one or more Principal Circuit nodes. Shunt arcs might complete alternate circuits within blocks.

A shunt can go either parallel or antiparallel (opposite) to the direction of the Principal Circuit. For example, the shunt from node 7 to node 5 parallels the path between nodes 7, 6, and 5. If we were to tear arcs into or out of node 6, there would still be one complete circuit in the graph. We must tear arcs from node 4 to node 3 or from node 3 to node 2.

- GO-BYS

Structural considerations alone, it is usually good practice to choose tears to create the smallest-size blocks along DSM spreadsheets' diagonal. This means fewer elements would be involved in circuits (shorter iteration

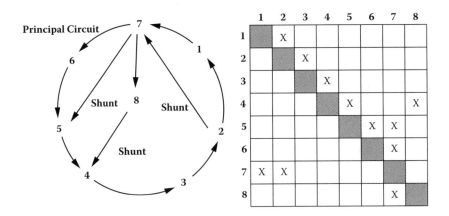

FIGURE 20.1
Graph containing shunts and its equivalent DSM spreadsheet.

times, smaller team sizes, clearer colocation requirements, facilitated and frequent communication, etc.).

The *node* on the Principal Circuit having the most structural reordering power is the node having the smallest number of shunt arcs that are parallel to (*go by*) this node. GO-BYs might complete circuits even if this node on the Principal Circuit were torn. The number of GO-BYs is probably the most important structural indicator of good places to tear a DSM spreadsheet.

- The node with the fewest or no GO-BYs has the greatest power for breaking apart blocks in the DSM spreadsheet.

21
Configuration Testing

For successful configuration testing, we need to know in what order to assemble individual software systems into integrated working configurations to test. So there will not be too many problems to diagnose, we pick enough systems to ensure adequate functionality. We add more systems until we have confidence in the overall system.

Building configuration tests this way can be time consuming and difficult. We might choose testing strategies that are easy to carry out but might be insufficient to ensure fully competent testing.

How can we be sure we have identified all critical dependencies? To find out, we applied a DSM spreadsheet to a purely structural analysis of a corporate-wide software system to double-check that our team had not missed any testing issues.

The Corporate Y2K (Year 2000) Remediation Plan was based on program content and usage but not on configuration and interconnections. The need was to understand both the global structure and the local dependencies among software threads and business systems.

- With a DSM spreadsheet, we completely identified and classified all the critical dependencies.

Our Y2K Program Office categorized all software systems according to their business impact:

Level **A**	Critical Business Failure
Level **B**	Major Disruption Failure
Level **C**	Nuisance Level Interrupt
Level **D**	Tolerable Malfunction

The **A**-level systems were identified as having a major business continuity impact. For example, billing systems, customer support systems, and employee safety systems were **A** level.

Yet, categorizing systems by criticality levels alone was insufficient for comprehensive configuration testing. If an **A**-level system received an input from a **D**-level system, it was important to check that interconnection, even though one of the interconnected systems by itself had a low criticality level.

Clusters with date information can pass data among themselves in circuits. Diagnosis is far more complex for a circuit because faulty data can appear in every system in the circuit, even though only one input/output (I/O) operation or system has an error. If a circuit exists, one solution is to tear the circuit so it appears as a linear chain of connected systems. Then, there is no risk that faulty data pass from the end of the chain back to its start. Once we are confident that a chain performed as intended, we reconnect and test the circuit.

The Y2K Program Office provided a Systems Interface Master Table listing of all software systems identified as critical to Y2K compliance. We sorted the 417 software systems into topological order. The new DSM spreadsheet consisted of 136 systems arranged sequentially (one at a time), followed by a cluster of 189 systems, followed by 92 systems arranged sequentially (one at a time).

Clusters are self-contained webs of dependent software systems. This means that every system within a cluster connects through a series of I/O exchanges with every other system within the cluster. The Principal Circuit was a circuit of 90 sequentially dependent systems, with the other 99 systems connected to this circuit as shunts or pair-wise appendages. Because they communicated with other systems through only one system on the Principal Circuit, they could be treated pair-wise independently. Based on structure, we excluded all these appendages.

We found the critical Principal Circuit of interconnected systems (in their sequential order) and broke the 189-system cluster into several smaller clusters as shown in Figure 21.1.

Understanding the interconnected structure of the software systems helped us establish clear priorities for configuration testing. Our review included both structural and business dependency analysis based on what specifically interconnecting these systems meant to the business. We pruned away those systems that did not require configuration testing. Our decisions were made either on our knowledge of business risk or on dependency structure.

Configuration Testing • 95

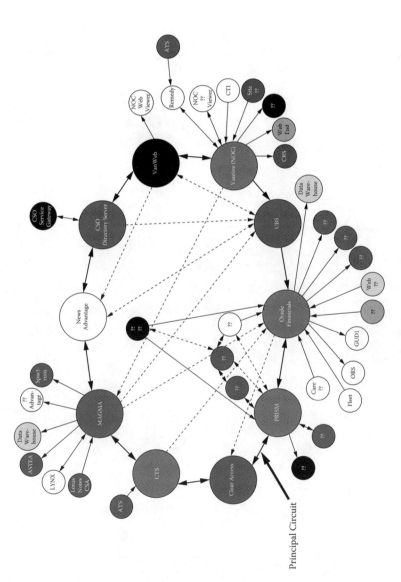

FIGURE 21.1
Y2K enterprise software diagram showing dependency paths identified by DSM spreadsheet analysis. Black, level **A**; dark gray, level **B**; light gray, level **C**; white, level **D**.

22

Control Dependencies

A *control dependency* is the critical, causal link between elements that determines or *controls* our situation. To illustrate and amplify, let us use a DSM spreadsheet to help us understand a nonbusiness, but important, social problem: inner-city crime and violence.

Don Steward provided this case to illustrate *control dependencies*. He has used DSM spreadsheets to study social problems. The DSM spreadsheet elements in Figure 22.1 are the issues that affect the commission of crimes and violence.

- *For emphasis, we purposely chose to create DSM spreadsheets by listing elements in Figure 22.1 in alphabetical order.*
- *For each element, we only identify its direct predecessor dependencies.*
- *In this way, we also demonstrate here how, even focusing at the microlevel, we can, after topological reordering and tearing a DSM spreadsheet, obtain the required macroview that both clarifies and illuminates.*

The torn and topologically reordered DSM spreadsheet of Figure 22.2 suggests the most important way to alleviate crime might be through improving inner-city residents' *Low Self-Respect*.

The three **A**s are the tears. These **A**s point to the *control dependency*: Low Self-Respect.

- *Low Self-Respect* is the social problem that needs to be solved.

Abuse & Violence in Family as Child	**1**																			
Bad Associates		**2**								X										
Commission of Crime and Violence		X	**3**	X					X	X		X	X		X	X	X	X	X	X
Deterrence-Quick Arrest & Punishment				**4**																
Education & Training					**5**															
Employability					X	**6**			X											
Employment						X	**7**	X												
Employment Opportunities								**8**												
Excessive Use of Alcohol & Drugs									**9**		X									
Low Income									X	**10**						X				
Low Self -Respect	X				X		X		X	X	**11**			X						X
Occupation of Time							X					**12**								
Peer Pressure to Crime & Violence		X			X								**13**							
Perception of Disadvantaged Status					X	X	X			X	X			**14**		X				
Police Security															**15**					
Poor Physical Environment												X				**16**				
Possession & Use of Guns											X		X				**17**	X		X
Society's Rules Benefit Others but Not Me														X				**18**		
TV Violence																			**19**	
Violence in Current Family	X								X		X								X	**20**

FIGURE 22.1
Crime-and-violence problem issues described. **(Note: DSM spreadsheet elements were entered alphabetically.)**

Abuse & Violence in Family as Child	**1**																			
Deterrence-Quick Arrest & Punishment		**4**																		
Education & Training			**5**																	
Employment Opportunities				**8**																
Police Security					**15**															
TV Violence						**19**														
Excessive Use of Alcohol & Drugs							**9**								**A**					
Employability			X				X	**6**												
Violence in Current Family	X					X	X		**20**						**A**					
Employment				X				X		**7**										
Occupation of Time										X	**12**									
Poor Physical Environment											X	**16**								
Low Income							X					X	**10**							
Perception of Disadvantaged Status			X					X		X		X	X	**14**	**A**					
Low Self-Respect	X		X				X		X	X			X	X	**11**					
Bad Associates													X			**2**				
Society's Rules Benefit Other but Not Me														X			**18**			
Peer Pressure to Crime & Violence			X													X		**13**		
Possession & Use of Guns									X						X		X	X	**17**	
Commission of Crime and Violence		X			X	X	X		X		X	X	X			X	X	X	X	**3**

FIGURE 22.2
Crime-and-violence problem DSM spreadsheet visualized after partitioning and tearing.

The three **A** marks above the diagonal identify the cluster of issues that must be addressed to reduce *Low Self-Respect*. The California Commission on Crime and Violence did come to a similar conclusion. They used the term *Self-Esteem*.

23

Breakthrough Thinking*

A DSM spreadsheet can help us extract new insights from a supposedly well-understood business process. The case described here—brake system design—is the major iterative portion of the automobile brake system design-and-testing process.

Automotive brake technology is mature. Engineers have considerable experience with brake system design. Yet, using structural analysis to determine which dependencies control the design process led to a significant breakthrough in their thinking—to a strategy that cut time and saved lots of money.

Drivers want automobiles with quiet, smooth brakes that do not require frequent service. To the design engineers, this means brake systems should have little or no brake squeal or brake pulsation and should have a long life. These problems are known respectively as noise, pulsation, and wear.

Prior to the DSM spreadsheet analysis, the brake system design engineers held these three problems (noise, pulsation, and wear) to be the controlling features of the design/test/redesign iteration problems that they were experiencing.

Some of the brake system design process was carried out using computer simulations. However, their only design verification strategy was to design, build a *prototype*, and *road test*.

Failing to identify *all* the dominant design parameters or design decisions usually results in the need to redo all or some of the design effort. Any negative results or needed refinements required one or more iterative scenarios. A new approach was needed for this complex and too-familiar problem.

---

* Adapted from Black, Thomas, Fine, Charles, and Sachs, Emanuel. 1990. A Method for Systems Design Using Precedence Relationships: An Application to Automotive Brake Systems. *MIT Sloan Management Review* October, no. 3208-90.

- Engineers realized that their current approached worked—but could they do better?

All the parameter interactions as described by experienced brake system engineers were represented in a DSM spreadsheet. But, there appeared to be so many interactions that brake system performance could not be fully described by predictive analytical or simulation models. Iteration using predictive models would be relatively fast. However, because creating good predictive models seemed overwhelming, there still seemed to be many lengthy iterations required in the brake system design process. These iterations included costly and time-consuming test track experiments.

This is *not so anymore*. Taking a close look at their DSM spreadsheet modeling the brake system design process, they found the 28 tightly coupled, iterative block of design elements shown in Figure 23.1. In the complete DSM spreadsheet, it was preceded by and followed by up to 200 additional largely sequential design parameters. In total, the complete DSM spreadsheet model contained over 400 additional elements.

They continued the process.

After partitioning and tearing the block in Figure 23.1, they saw that if they choose materials for the rotor and front linings, their coupled design problem would break into two separate design problems: kinematics (applying pressure) and thermodynamics (cooling). This DSM spreadsheet is shown in Figure 23.2. The partitioned and torn DSM spreadsheet suggested a radically different engineering approach could be possible:

- Computer modeling and simulations.

Previously, engineers only used software to analyze brake system kinematics. They used expensive and time-consuming test track testing for thermodynamics issues. After using a DSM spreadsheet, they realized the enormous expense and time required by construction of prototype brake systems followed by testing could be replaced if they had the thermodynamics (cooling) computer modeling and simulations.

- So, they chose to invest in modeling and simulation software.

Breakthrough Thinking • 103

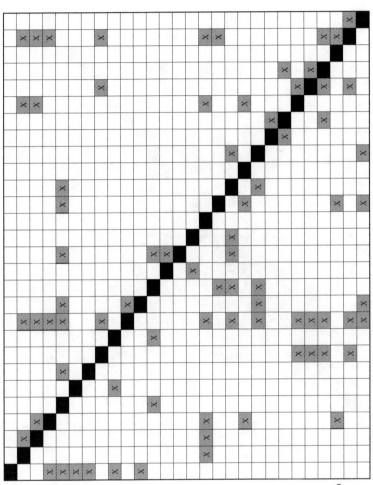

FIGURE 23.1
Brake system DSM spreadsheet before analysis.

104 • *Mastering Complexity*

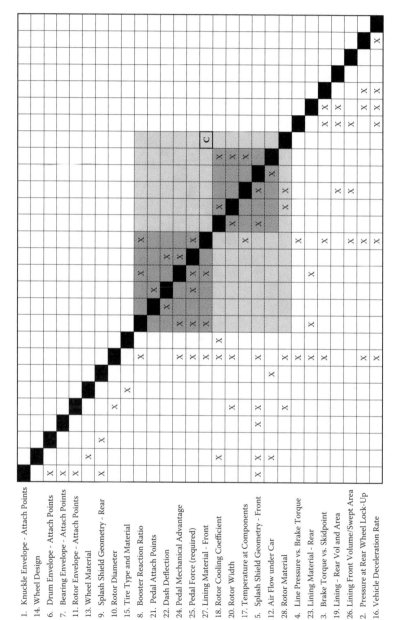

FIGURE 23.2
Brake system DSM spreadsheet after first partitioning and tearing.

Once they chose materials for the rotor and front lining material, two groups of engineers could work independently, in parallel, to shorten design time.

They even took their DSM spreadsheet analysis one step further. They tore one of the smaller blocks shown in Figure 23.2 to enable the mechanical design team to divide the work into two separate teams. This DSM spreadsheet is shown in Figure 23.3.

106 • *Mastering Complexity*

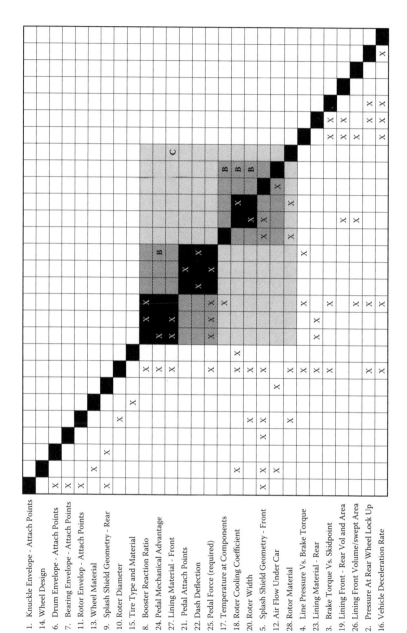

FIGURE 23.3
Brake system DSM spreadsheet after second partitioning and tearing.

Section 5

Putting DSM Spreadsheets to Work

Progress is made by lazy men looking for easier ways to do things.

—**Robert Heinlein**

24

Understanding Structure Essentials

Previous chapters provided examples of how, by making their dependency structure explicit and understandable, DSM spreadsheets can help us design work processes and systems that better support our business requirements. The approach in this book uses a square spreadsheet, the DSM, to visualize structure. We used this knowledge to manipulate business processes.

Businesses consist of networks of linkages and mutual dependencies. These dependencies determine how our business performs. To achieve a common purpose, we must understand clearly what this essential structure means and how the elements of our business interact. With DSM spreadsheets, we can show clearly how individuals are connected. We can understand how their relationships are dependent. More important, we systematically revise and optimize these dependencies one at a time or as a whole.

As we interact with more people and with more of our organization, we come to see our actions cannot be considered independently. Dependency structure provides needed visibility. If we can see the impact of what we do, we can see how the success or failure of the enterprise depends on us.

- We can see both the forest and the trees.

Businesses are usually organized to facilitate solving what is believed to be the principal problem pattern for the organization. But, organizations do not face just one problem that can be solved successfully with one optimum organization. We must deal with many types of problems that require many different problem-solving structures.

- Use a DSM spreadsheet to take advantage of a problem's intrinsic structural constraints to devise more effective solutions.

A DSM spreadsheet could help your organization improve its management practices and change.
Good luck and the best of success.

Section 6

Tools

Let all things be done correctly and in order.

—**Corinthians**

25
Microsoft Excel®-Based Free Software

DSM_Program_V2.1 is a Microsoft Excel *MACRO* created by Massachusetts Institute of Technology (MIT) Sloan School of Management Professor Steven Eppinger's students to handle most of the DSM spreadsheet analysis operations discussed in this book. It is free to download (http://www.dsmweb.org/en/dsm-tools/research-tools/excel-MACROs-for-partitioning.html).

- HOW TO INSTALL

When you first open the MACRO, Microsoft Excel asks you to **ENABLE** MACROs, as shown in Figure 25.1.

After you **ENABLE**, a DSM spreadsheet menu is added to your Microsoft Excel menu ribbon as shown in Figure 25.2. You will also see a default 20-element DSM spreadsheet already implemented as a Microsoft Excel file on the computer screen. You can start modifying it or you can delete all the elements and add new ones in a new *Elements Info* tab. Just do not forget to update your DSM spreadsheet every time you apply modifications.

- HOW TO USE

The *Elements Info* tab is your and the program's reference for all element information, including every element's full name and description.

The *Elements Info* tab is the only tab in which you can make modifications, and you need to make manual modifications.

- DO NOT ADD OR DELETE DSM SPREADSHEET ELEMENTS IN OTHER TABS.

114 • *Mastering Complexity*

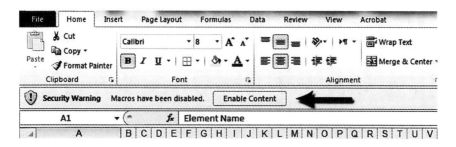

FIGURE 25.1
Enable macro content button.

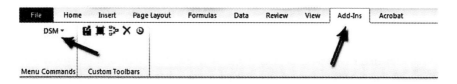

FIGURE 25.2
Macro is now added to Microsoft Excel menu ribbon.

They work automatically based on the information you provide in this tab.

CUTTING, COPYING, and *PASTING* cells in the DSM spreadsheet tab can cause problems. Use the new *Elements Info* tab to minimize such problems. Adding new elements or deleting, cutting, and pasting current elements is possible within the *Elements Info* tab. Modify any element in the *Elements Info* tab, hit the *Update DSM* button, and the DSM spreadsheet will be updated accordingly.

Currently, only binary DSM spreadsheets can be modeled. In spreadsheet cells, enter a **1** for an existing dependency and leave blank for none.

A *Swapping* command is used to interchange the positions of two elements with each other.

An *Insert* command allows the user to insert any number of new elements anywhere in the DSM spreadsheet.

A *Delete* command allows the user to delete a range of DSM spreadsheet elements from anywhere in the DSM spreadsheet.

A *Dependence Report* tab and a corresponding menu can be used to generate a dependency report for each DSM spreadsheet element. This helps tracking dependencies more easily, especially in a large DSM spreadsheet.

- Only one DSM spreadsheet file should be opened at a time. The MACRO code becomes mixed up when you open more than one file simultaneously.
- When an unexpected error pops up, it is usually best to close the file without saving and to reopen it.

The MIT Microsoft Excel MACRO is implemented using Microsoft Visual Basic, based on 32-bit Microsoft Office 2010. The MACRO was not tested on other platforms, but it should not have any problems. If you want to edit the MACRO or browse the code, go to the *View* menu tab in Microsoft Excel and select *Macros*; pick this MACRO and hit **EDIT**. Microsoft Visual Basic will open. Because this is a Microsoft Excel file supported by a MACRO (and not a fully programmed application), making it bulletproof against user manipulations is difficult. That is why the user must be careful when manually editing its contents.

26
Problematics PSM 32

Software to create, manipulate, and understand DSM spreadsheets is available from Don Steward's own website (http://www.problematics.com). Currently, his software is being used successfully in many business applications at companies large and small.

The *graph theory* methods used by PSM 32 are discussed at length in Donald Steward's book, *Systems Analysis and Management: Structure, Strategy and Design* (New York: Petrocelli, 1981). Here, we give a much-shortened description of how they work. We describe the procedure PSM 32 uses in terms of its equivalent graph. For a more detailed discussion, visit the website (http://www.problematics.com).

Analysis occurs in two stages: partitioning and tearing. Ordering requires we first identify the elements in circuits and confine them within DSM spreadsheet blocks. After partitioning, cells will have marks below the diagonal with possible exception of one or more square blocks symmetrically astride the diagonal. Any blocks are in topological order with the balance of DSM spreadsheet cells. Once we have partitioned DSM spreadsheets, tearing is used to reorder elements within blocks.

- PARTITIONING

Partitioning is done by path searching. PSM 32 software makes this all *hidden* to us—automatic and very fast. Here is how PSM 32 does it:

Step 1: PSM 32 first finds all the blocks and their Principal Circuits. Starting with the first row in the DSM spreadsheet, PSM 32 chooses any immediate predecessor element. A mark in a cell of this row means the element represented in the column is an immediate predecessor. PSM 32 then finds an immediate predecessor of this predecessor and so on. This continues until some element has been encountered twice. A closed path (circuit) has been traversed from the initial element back to itself.

Step 2: If PSM 32 finds a closed path, PSM 32 examines the circuit to find its Principal Circuit. PSM 32 labels the first circuit it found tentatively as the Principal Circuit. PSM 32 tests this assignment as it traces all shunts (counts their arc lengths) associated with this tentative Principal Circuit. Each time it finds a shunt path longer than the tentative Principal Circuit, PSM 32 interchanges this shunt path with the path in the tentative Principal Circuit and relabels the two. This procedure continues until all the shunts have been examined.

Step 3: PSM 32 takes all the nodes in the block and collapses them into one node so they can be treated jointly as one node during further searches for circuits. This one node, chosen arbitrarily, represents the set of the elements in the block. This representative node might have an arc to or from an element in the remaining graph.

Step 4: PSM 32 repeats this process until every element in the DSM spreadsheet has been examined. If there are no more circuits, we are done. Otherwise, PSM 32 continues.

Once this process is over, PSM 32 automatically generates a new DSM spreadsheet.

- The blocks are expanded to show all the elements within each block. Blocks reappear symmetrically astride the diagonal.
- Outside any blocks, all the marks in the DSM spreadsheet occur below its diagonal.
- Within blocks, there are marks both above and below the diagonal.

The rows and columns of the initial DSM spreadsheet are now reordered topologically.

- TEARING

We next use PSM 32 to topologically order elements within blocks internally. Our goal is to manipulate the DSM spreadsheet so it reveals the clearest understanding. *We work with PSM 32 interactively,* choosing tears that make sense to us. This is a trial-and-error process.

To control how we reorder the nodes during tearing, we replace simple marks with *Tearing-Level Numbers* on a 10-point scale of **0** through **9**. Here, **0** denotes the strongest dependency between two elements, **9** the weakest.

The tearing process starts with the weakest dependency (highest *Tearing-Level Number*) working from the highest assigned *Tearing-Level Number* toward the lowest (**9** toward **0**).

Tearing-Level Numbers denote arcs we remove (*tear*) to reveal underlying block structure. By assigning *Tearing-Level Numbers*, we have control over reordering. PSM 32 successively removes from consideration all nodes having the highest assigned *Tearing-Level Number* one tearing-level number at a time. PSM 32 does this for all blocks.

- Topological order is maintained.

Once we are done tearing the DSM spreadsheet, PSM 32 restores any nodes temporarily removed. The *Tearing-Level Number* is used to denote any cell having a dependency that was *torn*.

When we start out, we might find that we have to make many tears before we begin to see any emerging structure. Does our tear make sense?

We could choose *Tearing-Level Numbers* based on their dependency strength or on our confidence we know their impact. If their effect is small, we assign a high number so we can see more easily structures that involve strong dependencies. We could use the same 10-point scale as topological ordering. But, we prefer to use a simple 3-point scale of **A** for *strong*, **B** for *moderate*, and **C** for *weak*.

- GO-BYS

A low number of GO-BYs indicates tearing this node has a lot of leverage in tearing circuits because few shunts pass it by. There is a greater chance that the most circuits will be broken if we were to tear arcs into or out of this node. The smaller the number of GO-BYs, the more likely the block will shrink significantly when this node is torn and the DSM spreadsheet repartitioned.

PSM 32 helps us by identifying and characterizing GO-BYs in its *Tearing Advice Table*. GO-BYs are probably the most important structural indicator of a good place to tear DSM spreadsheets. The *Tearing Advice Table* shows the number of GO-BY shunts that would bypass a particular node if we were to tear the Principal Circuit here.

The number of GO-BYs suggests the power of a node in tearing the Principal Circuit. Some GO-BYs are more equal than others. When we see a small number of GO-BYs, we go for it. If one or more nodes have equal GO-BYs, we choose the node with the largest number of arcs that would be broken. Again, the PSM 32 *Tearing Advice Table* gives the needed information.

We also see if our tears might make good places to use assumptions. PSM 32 does not know this, but possibly we do. We usually need several tries before we obtain a set of tears with which we feel comfortable. PSM 32 can compute the results of our various choices quickly, so we can afford to try a number of choices.

27

Lattix Architect and Lattix Analyst

Lattix (http://lattix.com/) offers several DSM spreadsheet analysis and management software products to accomplish complexity management. Lattix Architect and Lattix Analyst are desktop applications that enable you to create dependency models of your systems, including applications, databases, services, and configuration files. With Lattix Architect, you can analyze your architecture in detail, edit the structure to create what-if and should-be architectures, and create design rules to formalize and communicate that architecture to your entire development organization.

Lattix also pioneered using the DSM spreadsheet approach for automated enforcement of software architectures and now supports analysis of enterprise systems, including processes, requirements, tests, hardware, infrastructure, and organizations, providing powerful visualization and change impact analysis across the entire system. The Lattix System includes many modules, tools, and features to extract dependency information and create DSM spreadsheets from UML/SysML (Unified Modeling Language/Systems Modeling Language) models, software code bases, SQL (Structured Query Language), databases, files, and other tools.

Index

A

actions, 4, 15, *see also* Tasks
advice, tearing, *see* Tearing
aircraft manufacturer example, 43–49
alignment, organization, 33
antiparallel shunts, 91
arbitrary partitioning order, 86
arcs
    defined, 9
    planning, 17
    process model, 13
    shunts, 91
    tearing circuits, 87
artificial circuits, 82
assumptions
    catching, 43
    making, 63–65
    manipulation, 67–70
    reviews, 25, 27
    risk management, 71–74
    tearing, 88
automotive examples
    brake system design, 101–106
    electric car, 75–78
    processes, 37–42

B

backward planning, 67
billing systems example, 94
blocks
    partitioning, 85
    planning, 17
    reviews, 27
    topological order, 81
bottom-up process, 53
brake system design example, 101–106
breaking circuits apart, 11, *see also* Tearing
breakthrough thinking, 101–106
building dependency maps
    assumptions, 63–65, 67–70
    dependency, 53–58
    projects and structure, 59–61
    risk management, 71–74
    unwrapping circuits, 75–78
business case
    design rules, 35
    elements identification, 31–32
    mapping dependencies, 32
    organizational alignment, 34
    redesigned, 33
business processes and structures
    complexity representation, 43–49
    continuity, 94
    cross-organizational collaboration, 37–42
    topological order, 81
    visualizing a business, 31–35

C

cascade, triggering, 5
cells, 13
circuits, *see also* Tearing
    breaking apart, 11
    defined, 10–11
    planning, 17
    reordering, 11
    topological order, 81–82
    unwrapping, 11, 75–78
closed paths, *see* Circuits
clusters, 94
columns
    assumptions, 63
    process model, 13, 16
    tearing, 88
commercial building project example, 67–68
commitment cycle, 24
communications
    cross-organizational collaboration, 37, 41
    direct dependencies, 57
complexity representation, 43–49

concurrency, 26, 43, *see also* Parallel paths and parallelism
configuration testing, 93–95
consistency, 57
constraints, 37, 83
control dependencies, 97–100
COPYING command, 114
Corinthians (biblical), 111
creation-fulfillment process cycle, 24
criticality
    assumptions, 63
    categorizing systems, 94
    cross-organizational collaboration, 38
    tearing, 88
cross-organizational collaboration, 37–42
customer commitment cycle, 24
customer support systems example, 94
CUTTING command, 114
cycle time reduction
    aircraft manufacturer example, 43
    shoes example, 20
    work structure, 23–24

D

decomposition, *see* Tearing
Delete command, 114
dependencies
    business units, 31–32
    classification, 67–68
    control dependencies, 97–100
    direct, 55–58
    management, 4, 23
    mapping, 32
    number of, 55
    overview, 53
    process model, 14–16
    shunts, 91
    simultaneous satisfaction, 23
    types, 54–55
    workflow driven, 24
Dependency Report tab, 114
dependency structure matrix (DSM) spreadsheets
    overview, *ix–x*
    process model, 13–16
    structure essentials, 109–110
descriptive statements, 75–76
design elements, 23
design process, 32–33
design rules, 33–35
direct dependencies, 15–16, 55–58
directed graphs, 9, 13

E

EDIT command, 115
electric car example, 75–78
elements
    breakthrough thinking, 102
    defined, 9
    dependencies, 53
    design process, 32–33
    direct dependencies, 56
    identification, 31
    partitioning, 85
    process model, 13
    reordering, 63–64
    structure, 109
    topological order, 81
Elements Info tab, 113–114
employee safety systems example, 94
ENABLE command, 113
Eppinger, Steven, 113
errors, *MACRO,* 115
essentials of structure, *see* Structure
examples
    aircraft manufacturer, 43–49
    automobile company, 37–42
    brake system design, 101–106
    commercial building project, 67–68
    electric car, 75–78
    inner-city crime/violence, 97–100
    inventory management system replacement, 72–74
    microfilm manufacturer, 59–61
    shoes, 19–21
    software company, 24
    Y2K compliance, 5–7

F

fast-tracking, 68, 70
feedback
    projects and structure, 59
    reordering elements, 63–64
    reviews, 27
    tearing, 88

feedforward, 59
feeds, 53, *see also* Owing others
forest and trees analogy, 109
foundation example, 68
functional dependencies, 54–55
functionally oriented approach, 45–47

G

Gantt charts, 59
Get Shoes action, 19–21
Get Socks action, 19–21
global view, 15
GO-BYs
    PSM 32, 119–120
    structural components, 91–92
graphs
    structural model, 9–10
    theory methods, 117

H

handoffs, 37, 38
hardwire, task order, 23
Heinlein, Robert, 107

I

indirect dependencies, 55
information, flows and exchanges
    collaboration, 23
    cross-organizational collaboration, 37
    direct dependencies, 57
information dependencies, 54
inner-city crime/violence example, 97–100
Insert command, 114
Inspect Shoes action, 19–20
inventory management system replacement example, 72–74
iteration
    assumptions, 63
    breakthrough thinking, 102
    business visualization, 33
    non-value added, 18
    planning, 17
    risk management, 71–72
    sequence issues, 23
    shoes example, 19
    unwrapping circuits, 77

K

kinematics, 102

L

Lattix software, 121
local view, 15
logical circuit, *see* Circuits
logical structure, exposing
    breakthrough thinking, 101–106
    configuration testing, 93–95
    control dependencies, 97–100
    partitioning, 85–86
    structural components, 91–92
    tearing circuits, 87–89
    topological order, 81–83
logistics view, 4, 23
loops, *see* Circuits
lowercase letters, 75
low self-respect example, 97–-100

M

MACRO (Excel software), 113–115
mainframe system replacement example, 72–74
making assumptions, 63–65
manipulation, assumptions, 67–70
manufacturability, design rules, 35
manufacturing capacity
    design rules, 35
    elements identification, 31–32
    organizational alignment, 34
    redesigned, 33
mapping dependencies, 32
market testing
    design rules, 35
    elements identification, 31–32
    mapping dependencies, 32
    organizational alignment, 34
    redesigned, 33
Massachusetts Institute of Technology (MIT), 113
materials procurement example, 68
microfilm manufacturer example, 59–61
Microsoft Excel-based free software, 113–115
Microsoft Office software, 115

Milne, A.A., *xv*
mixed dependencies, 55

N

needs, 14–16, 53
Niebuhr, Reinhold, 79
night batch-update example, 72–74
nodes
    defined, 9–10
    GO-BYs, 92
    planning, 17
    process model, 13
    tearing circuits, 87
non-value added iteration loops, 18, 24

O

ordering, *see also* Rearrangement
    hardwired, 23
    non-value added iteration loops, 24
    out-of-order loops, 24
    projects and structure, 59
organizational alignment, 33
organization chart, 3
overbuilding, 68
owing others, 14–15, *see also* Feeds

P

pair-wise appendages, 94, *see also* Shunts
parallel paths and parallelism, *see also* Concurrency
    aircraft manufacturer example, 48
    brake system example, 105
    defined, 10
    partitioning, 85
    planning, 17
    shoes example, 20–21
    shunts, 91
    unwrapping circuits, 77
partitioning
    analysis, 117
    brake system example, 104, 106
    breakthrough thinking, 102
    commercial building example, 69
    crime-and-violence problem, 99
    logical structure, 85–86
    PSM 32, 117–118

PASTING command, 114
paths, 9, *see also* Parallel paths and parallelism
PERT charts, 59
planning
    handoffs, 37
    working back, 67
    work structure, 17
predecessors, 13
preferential dependencies, 54
prescriptive processes, 75–76
Principal Circuits, 91–92
problem-solving advice, 89
product concept
    design rules, 34–35
    elements identification, 31–32
    mapping dependencies, 32
    organizational alignment, 34
    redesigned, 33
product development process, 31
product development tasks, 39–40
program evaluation and review technique (PERT) charts, 59
projects and structure, 59–61
Project Steering Team, 41–42
PSM 32 software, 117–120
Put on Shoes action, 19–20
Put on Socks action, 19–20

Q

qualitative and quantitative data, 56

R

rearrangement, redesigning, and reordering
    aircraft manufacturer example, 48–49
    assumptions, 63–64, 67
    business visualization, 33
    circuits, 11
    feedback loops, 74
    partitioning, 85
    process model, 13
    product development tasks, 40
    review process, 24
    topological order, 82
resource dependencies, 54
responsibilities

cross-organizational collaboration, 37, 38, 41, 42
direct dependencies, 57
product development tasks, 40
reviews
unwrapping circuits, 77
work structure, 25–27
rework, 24, 63
risk management
accepting, 73
assumptions, 64, 67–68
overview, 71–74
rows
assumptions, 63
process model, 13, 16
tearing, 88

S

scheduling
aircraft manufacturer example, 47, 49
assumptions, 63
circuits, 11
Sears example, 4
self-contained webs, 94
self-esteem, 100, *see also* Inner-city crime/violence example
semantics, 3–4
sequence
aircraft manufacturer example, 46, 48
assumptions, 68
could-be, 68
node arrangement, 10
topological order, 81
shoes example, 19–21
shunts, 91, 94
simultaneous satisfaction, 23
Sloan School of Management, 113
software, *see also* Tools
investment, 102
tearing suggestions, 88, 119–120
software company example, 24
solutions constraints, 37
sourcing strategy
design rules, 35
elements identification, 31–32
organizational alignment, 34
redesigned, 33
square blocks, 17

stability, waterfall process, 72
Steward, Don, 97, 117
structure
components, 91–92
cross-organizational collaboration, 37
cycle time reduction, 23
defined, 3–4
dependencies management, 4
essentials, 109–110
model, 9–11
overview, 3–5
subdividing blocks, 27
successors, 13
Swapping command, 114
system-oriented approach, 44
Systems Analysis and Management: Structure, Strategy and Design, 117
Systems Interface Master Table listing, 94

T

tasks and task management, *see also* Actions
arbitrary, 20
cross-organizational collaboration, 37–38, 41
cycle time reduction, 23
dependencies, 54
design process, 32–33
hardwired, 23
recasting our view, 4
reviews, 27
tearing, *see also* Circuits
advice, 88, 119–120
analysis, 117
brake system example, 104, 106
breakthrough thinking, 102
circuits, 76, 94
control dependencies, 97
crime-and-violence problem, 99
GO-BYs, 91–92
logical structure, 87–89
partitioning, 85
problem-solving advice, 89
PSM 32, 117, 118–119
software suggestions, 88, 119–120
topological order, 83
Tearing Advice Table, 119–120

technical feasibility
    design rules, 34–35
    elements identification, 31–32
    organizational alignment, 34
    redesigned, 33
thermodynamics, 102
"Tiger" (manager), 41–42
tools
    Lattix Architect and Lattix Analyst, 121
    Microsoft Excel-based free software, 113–115
    partitioning, 85
    PSM 32 software, 117–120
top-down approach, 53
topological order
    control dependencies, 97
    logical structure, 81–83
    partitioning, 85–86
    PSM 32, 118, 119
    tearing circuits, 87
trade-offs, 37, 38
trees and forest analogy, 109

U

uncertain dependencies, 54
UNIX system example, 72–74
unwrapping circuits, 11, 75–78, *see also* Circuits
Update DSM button, 114
uppercase letters, 75

V

verification, 25–27
View menu tab, 115
visibility, 4, 9
Visual Basic, 115
visualization, business, 31–35

W

Wang, An, 29
warehouse inventory management system example, 72–74
waterfall process, 72
weak dependencies, 67, *see also* Dependencies
Whitehead, Alfred North, 1
work structure
    cycle time reduction, 23–24
    DSM spreadsheet process model, 13–16
    overview, 3–5
    plans, 17
    reviews, 25–27
    shoes example, 19–21
    structural model, 9–11

Y

Y2K (year 2000) compliance example, 5–7
Y2K (year 2000) remediation plan example, 93–95

About the Author

Stephen Denker is the managing partner of Business Process Architects. He has more than 30 years of business and industrial experience. Business Process Architects was formed to help companies design and simplify their core work processes. Its objective is to strengthen competitive performance by developing integrated improvements. Its services and tool set enable clients to articulate, and then verify, that any new processes or systems better support business requirements.